Biocontrol
of
Plant Diseases

T0141107

Volume II

Editors

K. G. Mukerji, Ph.D.
Professor, Department of Botany
University of Delhi
Delhi, India

K. L. Garg
Mycology Laboratory
Department of Botany
University of Delhi
Delhi, India

CRC Press
Taylor & Francis Group
Boca Raton London New York

CRC Press is an imprint of the
Taylor & Francis Group, an **informa** business

CRC Press
Taylor & Francis Group
6000 Broken Sound Parkway NW, Suite 300
Boca Raton, FL 33487-2742

Reissued 2019 by CRC Press

© 1988 by Taylor & Francis Group, LLC
CRC Press is an imprint of Taylor & Francis Group, an Informa business

No claim to original U.S. Government works

A Library of Congress record exists under LC control number:

Publisher's Note
The publisher has gone to great lengths to ensure the quality of this reprint but points out that some imperfections in the original copies may be apparent.

Disclaimer
The publisher has made every effort to trace copyright holders and welcomes correspondence from those they have been unable to contact.

ISBN 13: 978-0-367-26280-8 (hbk)
ISBN 13: 978-0-367-26281-5 (pbk)
ISBN 13: 978-0-429-29234-7 (ebk)

Visit the Taylor & Francis Web site at http://www.taylorandfrancis.com and the
CRC Press Web site at http://www.crcpress.com

PREFACE

In the last few years research interest in biological control of plant diseases has become extremely active. One could easily notice this trend by following the number of publications in several reputable journals in plant pathology, mycology, and microbial ecology, where at least three to four hundred papers appear on this subject every year. One of the main reasons for this burst of research activity is the hazardous impact of various fungicides (pesticides) and other agrochemicals on the ecosystem. Therefore, it has become essential to do more work on the biological control of diseases and to avoid the use of fungicides and other chemicals, considering the ecological damage which may result. We were convinced that there was a need for a book which could summarize the existing work on the biocontrol of plant diseases. It is our good fortune that many of the leading researchers, who are actively studying different aspects of this problem, became enthusiastic and agreed to contribute the reviews which form this book.

Under natural and normal conditions the microbes on the plant surface are in a state of dynamic equilibrium. Interactions between pathogens and nonpathogens, and also amongst themselves, are constantly in process. When a pathogen is established on its host it is evident that it is more or less safe from the influence of a multitude of other organisms. When the spores of a pathogen fall on a fresh host surface they are fully exposed to the influence of many other microbes which are natural inhabitants and already present there. They can live alongside others without causing any noticeable effect on their behavior or they may be suppressed. Plants, in a natural (wild) state, possess a relatively stable biological balance of the microbes on their surface. An alien organism introduced into an area in which it has no natural enemies may increase in number to such an extent that the resident population is unable to redress the balance. This presents a problem which is best solved by introducing a second organism chosen with great care, and taking into account many necessary safeguards which will selectively attack the unwanted species. This is "biological control" (biocontrol). Almost any process, occurring naturally or done artificially, which affects the relationship between organisms in such a way that the natural biological balance is restored, can be regarded as biocontrol.

Pathogens are specific in their attack to a particular organ surface of the host. They may enter through the root from soil, or via air in the aerial parts of the plants, i.e., the leaf, vegetative shoot, bud, or flowers. During the last 15 to 20 years enormous amounts of work have been done on the biological control of root and soil-borne diseases and also on diseases of the aerial parts of plants. Different types of pathogens have been controlled by either antagonistic saprophytes or hyperparasites. This has also been done by artificially inoculating the biocontrol agent, or by increasing its number by amending the soil with nutrients, moisture content, pH, etc., and by crop rotation. Soil solarization is a new approach to soil disinfection and is based on the solar heating of the soil during the hot season, thereby increasing the temperature and killing the pathogens. Crop protection can also be done by inducing resistance in a host by a biocontrol agent. Mutation of the pathogens due to insertions, base pair substitutions, or deletions can be made on isolated genes, and these modified genes can be reintroduced to recipient strains to cause site-specific mutagenesis of these recipients. These strains will be very effective biological control agents.

Each chapter is devoted to a separate aspect of biocontrol of plant diseases, and in general the chapters are arranged in order of increasing technical complexity. Since these chapters have been written by independent authors there is the possibility of a slight overlap or repetition of certain statements, but this is difficult to avoid with this sort of organizational structure. One of the chapters deals with the biocontrol of weeds, which some may feel out of place, but it is important to have it as weeds cause enormous losses to agriculture.

It is our hope that this book will be useful to all students and researchers in plant pathology, microbial ecology, and mycology.

We are indebted to Dr. Nalini Nigam and Dr. Geeta Saxena for help in various ways.

31st July, 1986

K. G. Mukerji
K. L. Garg

EDITORS

K. G. Mukerji, M.Sc., Ph.D., is a Professor and presently the Head of the Department of Botany, University of Delhi, India.

Professor Mukerji did his Masters degree in Botany at Lucknow University, Lucknow, in 1955 and obtained his Ph.D. in Botany at the same University in 1962. He is internationally known for his work on fungal taxonomy and microbial ecology. He has worked, or visited, in most of the important laboratories of the world concerned with Mycology and Plant Pathology.

Professor Mukerji is a member of most of the Societies or Associations concerning Mycology, Plant Pathology, and Microbial Ecology. In some of these societies he has also held important offices.

Professor Mukerji has presented over 20 invited lectures at International Meetings, over 30 invited lectures at National Meetings, and approximately 150 guest lectures at Universities and Institutes. He has published more than 220 research papers. He has coauthored *Taxonomy of Indian Myxomycetes* and *Plant Diseases of India,* and edited *Progress in Microbial Ecology.* Presently he is editing an annual volume, *Frontiers in Applied Microbiology,* two volumes of which are already out. He is on the editorial board of several journals dealing with Mycology and Plant Pathology. His current major interests include microbial ecology, mycorrhizal technology, and biocontrol of plant diseases.

Professor Mukerji is a distinguished mycologist and microbial ecologist and he is well respected for his research contributions all over the world.

K. L. Garg, M.Sc., Ph.D., is presently a Junior Mycologist at the National Research Laboratory for Conservation of Cultural Property, Lucknow, U.P., India. After completing his Ph.D. in Botany from Lucknow University in 1982, he joined Professor K. G. Mukerji at Delhi University as a Post Doctoral Fellow. He has published more than 25 research papers.

CONTRIBUTORS, VOLUME I

Ashok Aggarwal, Ph.D.
Lecturer
Botany Department
Kurukshetra University
Kurukshetra, Haryana, India

K. R. Aneja, Ph.D.
Lecturer
Botany Department
Kurukshetra University
Kurukshetra, Haryana, India

Basil M. Arif, Ph.D.
Research Scientist
Department of Virology
Forest Pest Management Institute
Sault Ste. Marie, Ontario, Canada

Abdul Ghaffar, Ph.D.
Professor
Department of Botany
University of Karachi
Karachi, Pakistan

A. K. Gupta, Ph.D.
Senior Research Fellow
Botany Department
Kurukshetra University
Kurukshetra, Haryana, India

S. Hasan, Ph.D., D.Sc.
Principal Research Scientist
Biological Control Unit
CSIRO
Montpellier, France

Peter Jamieson, M.Sc.
Research Associate
Forest Pest Management Institute
Canadian Forestry Service
Sault Ste. Marie, Ontario, Canada

Guy R. Knudsen, Ph.D.
Assistant Professor
Division of Plant Pathology
Department of Plant, Soil and
  Entomological Sciences
University of Idaho
Moscow, Idaho

R. S. Mehrotra, Ph.D.
Professor and Chairman
Department of Botany
Kurukshetra University
Kurukshetra, Haryana, India

Peter C. Mercer, Ph.D.
Faculty of Agriculture
Queen's University of Belfast
Belfast, Ireland

Nalini Nigam, Ph.D.
Research Associate
Department of Botany
University of Delhi
Delhi, India

Bharat Rai, Ph.D.
Professor
Department of Botany
Banaras Hindu University
Varanasi, India

Geeta Saxena, Ph.D.
Research Associate
Department of Botany
University of Delhi
Delhi, India

H. W. Spurr
Biocontrol Research Unit
U.S. Department of Agriculture
Oxford, North Carolina

Leif Sundheim, Ph.D.
Associate Professor
Department of Plant Pathology
Norwegian Plant Protection Institute
Agricultural University of Norway
AAS-NLH, Norway

Arne Tronsmo
Associate Professor
Department of Microbiology
Agricultural University of Norway
AAS-NLH, Norway

R. S. Upadhyay, Ph.D.
Lecturer
Department of Botany
Banaras Hindu University
Varanasi, India

Homer D. Wells, Ph.D.
Research Plant Pathologist
U.S. Department of Agriculture
Department of Plant Pathology
University of Georgia
Tifton, Georgia

CONTRIBUTORS, VOLUME II

Raul G. Cuero, Ph.D.
Microbiologist
Agricultural Research Station
U.S. Department of Agriculture
Southern Regional Research Center
New Orleans, Louisana

Jane L. Faull, Ph.D.
Lecturer
Department of Botany
Birkbeck College
London, England

D. W. Fulbright, Ph.D.
Associate Professor
Department of Botany and Plant Pathology
Michigan State University
East Lansing, Michigan

C. M. E. Garrett, Ph.D.
Plant Bacteriologist
Department of Plant Pathology
Institute of Horticultural Research
East Malling, Maidstone
Kent, England

Sally Westveer Garrod, M.S.
Department of Botany and Plant Pathology
Michigan State University
East Lansing, Michigan

W. Janisiewicz, Ph.D.
Research Plant Pathologist
Agricultural Research Station
Appalachian Fruit Research Station
U.S. Department of Agriculture
Kearneysville, West Virginia

Charles M. Kenerley, Ph.D.
Assistant Professor
Department of Plant Pathology and
 Microbiology
Texas A&M University
College Station, Texas

Gurdev S. Khush, Ph.D.
Principal Plant Breeder
Plant Breeding Department
International Rice Research Institute
Manila, Phillipines

Eivind B. Lillehoj, Ph.D.
Supervisory Microbiologist
Southern Regional Research Center
U.S. Department of Agriculture
New Orleans, Louisana

Alison Murray
Biosciences and Biotechnology Department
University of Strathclyde
Glasgow, Scotland

Cynthia P. Paul, M.S.
Graduate Assistant
Department of Botany and Plant Pathology
Michigan State University
East Lansing, Michigan

R. E. Pettit, Ph.D.
Associate Professor
Department of Plant Pathology
Texas A&M University
College Station, Texas

James E. Rahe
Department of Biological Sciences
Center for Pest Management
Simon Fraser University
Burnaby, British Columbia, Canada

K. V. Sankaran, Ph.D.
Scientist
Division of Forest Pathology
Kerala Forest Research Institute
Peechi, Kerala, India

Rudy J. Scheffer, Ph.D.
Phytopathologist
Willie Commelin Scholten
Phytopathological Laboratory
Baarn, Netherlands

Jyoti K. Sharma, Ph.D.
Scientist-in-Charge
Division of Forest Pathology
Kerala Forest Research Institute
Peechi, Kerala, India

Jagjit Singh, Ph.D.
Research Officer
Department of Biology
Birkbeck College
London, England

John E. Smith, Ph.D.
Professor
Biosciences and Biotechnology Department
University of Strathclyde
Glasgow, Scotland

James P. Stack, Ph.D.
Assistant Professor
Department of Plant Pathology and
  Microbiology
Texas A&M University
College Station, Texas

Gary A. Strobel, Ph.D.
Professor
Department of Plant Pathology
Montana State University
Bozeman, Montana

John Tuite, Ph.D.
Professor
Department of Botany and Plant Pathology
Purdue University
West Lafayette, Indiana

S. S. Virmani, Ph.D.
Plant Breeding Department
International Rice Research Institute
Manila, Phillipines

David M. Wilson, Ph.D.
Professor
Department of Plant Pathology
Coastal Plain Station
University of Georgia
Tifton, Georgia

Marcus Zuber, Ph.D.
Professor Emeritus
Department of Agronomy
University of Missouri
Columbia, Missouri

TABLE OF CONTENTS, VOLUME I

TABLE OF CONTENTS, VOLUME II

Chapter 1

BIOCONTROL OF RUST AND LEAF SPOT DISEASES *

J. K. Sharma and K. V. Sankaran

TABLE OF CONTENTS

*   KFRI Scientific Paper No. 133.

I. INTRODUCTION

Among the greatest hazards in crop production, unfavorable weather conditions, insect pests, and diseases are the main factors. Any one of them can upset the crop yields with catastrophic suddenness. Of all the plant diseases, foliar diseases, especially rusts and leaf spots, are the important group of diseases causing large-scale destruction of agricultural and horticultural crops. There are about 4000 species of rust fungi belonging to 100 genera, many of which are capable of causing widespread epidemics. A few notable examples are those of cereal rusts *(Puccinia graminis* f. sp. *tritici, P. graminis* f. sp. *secalis),* cedar rust of apple *(Gymnosporangium juniperi-virginianae),* and white pine blister rust *(Cronartium ribicola),* which are well known for causing economic losses. Cereal rusts are known to have changed the cropping pattern of different regions and food habits of the population. Wheat leaf rust created an economic disaster in 1916 in the U.S. and Canada. An even worse disaster struck in 1935 when about 100 million bushels of wheat were lost. In India, according to an estimate during 1971 to 1972, a brown rust *(Puccinia recondita)* epidemic in northwestern India resulted in a loss of 1.5 million tonnes of wheat.[79] Epidemics of the coffee rust compelled Sri Lanka to bring more land into tea plantations and to abandon coffee cultivation for years. This way coffee rust changed the entire economy of Sri Lanka in 1875 due to the widespread rust epidemic. Leaf spot diseases are equally significant in causing economic disasters. To name a few, late blight of potato *(Phytophthora infestans),* blast of rice *(Pyricularia oryzae),* Helminthosporium leaf spot of maize *(H. turcicum, H. maydis, H. carbonum),* and bean anthracnose *(Colletotrichum lindemuthianum)* are known to have affected crop yields drastically. The potato famine of Ireland in 1845, due to late blight, which gradually spread to the whole of the European continent, is the most cited example to indicate destruction and aftereffects of diseases on a large population. Rice blast disease, reported from 70 countries has affected up to 90% of the yield depending upon the part of the plant infected. In the 1950s in Florida, losses of winter-grown sweet corn due to leaf spot caused by *H. turcicum* amounted to 20 to 90%/field. These examples provide documented proof of the extent of damage caused by rusts and leaf spots. For combating these diseases successful measures of chemical control have been developed over the years. Though chemicals have played a significant role in maximizing the crop productivity, extensive, often excessive use of broad spectrum compounds, some of which are nondegradable, has resulted in a variety of harmful and undesirable effects not only on man and wildlife, but on the ecosystem as a whole. This way, relative short-term effective control of a disease makes this method less attractive. Control of plant diseases through genetic resistance of the host is often successful. But pathogens can and do mutate and, therefore, many of the promising resistant varieties succumb to disease in the field sooner or later. Besides, in tree diseases, both of these methods of disease control may not find a direct or immediate application due to the long period of the host's life. In this situation, with the increasing awareness of the problems and expense of conventional methods of disease control, including fungicides and costly and time-consuming breeding programs, biological control of plant pathogens has many attractions.

In many plant diseases the interaction of host, pathogen, and physical environment is readily apparent. The recognition of a fourth component, namely other organism(s) which affect the course of disease either by their direct influence on the pathogen or indirectly through their effect on the host or environment, provides the basis for our concept of biological control. The term "biological control" was first used by Smith[144] to signify the use of natural enemies to control insect pests. Stanford and Broadfoot[152] were the first to use the term "biological control" in plant pathology while working on the control of wheat take-all fungus. The scope and relevance of biological control has expanded over the decades. Debach[42] discussed the semantics of the term "biological control" and concluded that it is a natural phenomenon relating to those biotic agents that prevent the normal tendency of populations of organisms to grow in an exponential fashion and to the mechanisms by which such growth is prevented or regulated. Later definitions provided by Garrett[62] and Baker[12] agreed with Debach, but excluded man as a biotic agent. Snyder[146] takes a much broader view in his definition that "biological control relies largely upon an interruption of host-parasitic relationships through biological means . . . and may be accomplished by imparting resistance to the host . . . or by modifying the cultural practices of the crop so as to reduce the infection." Swell,[156] Baker and Cook,[11,37] and Cook[36] subscribe to this broader definition. Whatever the scope of definition one accepts, the original concept intended that in biological control the fourth component, namely "other living organisms," is brought in to play an important role regardless of its initiation. It is in this context that biological control is used here. Biological control by conventional methods such as cultural practices,[76,113] selection, and breeding for resistance[97] has been excluded and emphasis has been placed on microbiological biocontrol of plant pathogens.

Certain microorganisms have long been known to be associated with other pathogenic microorganisms in a variety of ways. The surface of aerial plant parts provides a habitat for epiphytic microorganisms.[23,46,140] They are also found in soil, on the roots of plants, and even within the plant tissues. To what extent the microorganisms are capable of direct or indirect biological control of rust and leaf spot diseases and how this capability can be exploited is the purpose of this review. An attempt has been made to bring together the scattered information on these two groups of diseases. The examples cited here may have some applications in the biocontrol, if not in practice then in principle, of rust and leaf spot diseases.

II. BIOLOGICAL CONTROL IN NATURE

In nature, microbial pathogens frequently cause epiphytotics, which help to maintain the balance of fungal populations.[72] Baker and Cook[11] point out that antagonists are likely to be acting continuously against potential pathogens, thus contributing to the absence of disease in a majority of the situations; biological control is thus already working in the field. How can the effectiveness of natural mechanisms of biological control be maximized with reference to the foliar environment? Two means of achieving this are apparent: (1) to allow natural control mechanisms to proceed as far as possible unhindered, or (2) to manipulate the system so as to give increased control. The former means could be aided by avoiding, as far as possible, the indiscriminate use of fungicides which destroy the epiphytic flora.[45,71] The latter approach could be used when a natural control mechanism existed but is shown to be only partly effective. However, it requires a much greater understanding of the mechanisms of action of possible controls and of interactions with other environmental factors. An awareness that biological control works in nature should provide the greatest stimulus to utilize the capacity of the natural microflora to bring about disease control. There are four possible ways by which microorganisms can affect the propagules, infection, and fructification of pathogenic microorganisms, and reduce or control the development of a disease. They are (1) hyperparasitism, (2) antagonism by naturally occurring and foreign microorganisms, (3) immunization or cross protection, and (4) hypovirulence. Now we shall deal with them separately and in detail with examples of various rust and leaf spot diseases.

A. Hyperparasitism

Hyperparasitism is the attack of a secondary parasite on a primary parasite. The term hyperparasite refers to the parasitism of one microorganism by another; a mycoparasite is a fungus parasitic on another fungus. Unlike antibiotic producing microorganisms, hyperparasites do not initiate their parasitic activity at a distance, but require an intimate association of the host and parasite. Some hyperparasites obtain nutrients from the living host causing little or no apparent harm; this is "balance" or biotrophic parasitism.[13] When the action of the parasite kills the host, the phenomenon is called necrotrophic hyperparasitism. It is the latter mode of action which is of most interest to those concerned with the biological control of plant pathogens. Hyperparasites of fungi were first observed in 1800s by mycologists interested in plant diseases. Most reports of hyperparasites suggest that they might be useful for biological control. The nature of hyperparasitism has been outlined by Barnett[13] and the topic reviewed by several authors.[24,27,40,43,46,87,100] Kuhlman[88] has discussed various attributes of hyperparasites providing biological control.

1. Rusts

Basidiomycetous fungi, particularly the rusts, have been frequently noted as hosts of other parasites. Since the rusts produce external fructifications as a secondary inoculum helping the disease spread, they are liable to be controlled more effectively by hyperparasites than other disease such as leaf spots. Both primary and secondary inoculum can be parasitized, thereby affecting the disease at the time of infection, and later, the subsequent spread.

a. Fungi

One of the earliest reported hyperparasites of rusts is *Darluca*. It occurs naturally on uredo- and teleutosori of numerous rust species. In early studies of asparagus rust, Smith[145] observed the hyperparasite, *Darluca filum*, growing on uredinia of *Puccinia asparagi*. Later, the parasitism of *Darluca* on hemlock rust was studied by Adams[2] who found that the mycelium of *D. filum* ramified throughout the pycnidial and aecial sori of *Peridermium peckii*, destroying the spores and disorganizing pycnial and aecial primordia. Keener[82] discovered biological specialization within *D. filum* on 16 macrocyclic and 3 microcyclic rusts. In greenhouse tests, isolates of *Darluca* were separated into 11 distinct groups. Susceptibility of the various rusts differed greatly; *Puccinia sorghii* was attacked by all isolates of *D. filum*, whereas *Puccinia poculiformis* was immune to all. The isolates of *Darluca* also differed in their virulence on the 19 differential rust hosts, but no relationship was found between the rust genera from which the isolates of *Darluca* were obtained and the genera on which they were parasitic. Pycnidia of *D. filum* were often embedded in the spore masses of the rust sori and were not visible macroscopically. The host range of *D. filum* has been appraised by McAlpine.[102] He found *Darluca* on 24% of the species of *Puccinia* and on species of many other rust genera in Australia.

Petrak[118] observed *D. filum* parasitizing *Uredo ravennae* on *Eriantheus ravennae*. Fedorinchik[54] isolated *D. filum* from mycelium and rust pustules of *Puccinia triticina* of wheat. He grew *Darluca* on sterilized pieces of carrot and potato and also on oats and corn seed to obtain inoculum for pathogenicity studies. Fedorinchik noted that when *Darluca* and *Puccinia* spores were inoculated simultaneously on wheat plants, the rust was completely destroyed in the mycelial stage. Bean[15] found that *D. filum* caused a significant reduction in the inoculum potential of race 6AF of oat stem rust (*Puccinia graminis* f. sp. *avenae*) and consequently a lower infection of the host. The availability of susceptible fungal tissue for a prolonged time seems a prerequisite for successful hyperparasitism, as was clearly demonstrated by comparing *D. filum* infection of rusts on evergreen oaks with that of deciduous oaks.[89]

D. filum has been found to parasitize uredinia of *Uromyces phaseoli* var. *typica* on bean, *Melampsora larici-populina* on poplar, and other uredinales attacking maize, sorghum, *Pennisetum purpureum*, and *Sorghum halepense*.[26,63,109,134] Under natural conditions parasitism on *U. phaseoli* was less than on the other species, but maize preceded bean, resulting in a high incidence of the rust and the parasitism by *D. filum*. It is also observed to attack urediniospores of *Puccinia ruelliae* and *Uromyces andropogonii-annulati* on *Dipteracanthus prostratus* and *Dicanthium annulatum,* respectively.[18]

The genus *Tuberculina* also attacks various genera of Uredinales. Smith[145] observed *Tuberculina persicina* growing upon asparagus rust. He also reported that a species of *Cladosporium* invaded and killed urediniospores of *Puccinia asparagi.* The association of *Cladosporium aecidiicola* and *T. persicina* with *Puccinia conspicua* on *Helenium hoopesii* was studied by Keener.[81] Meilke[107] investigated the parasitism of *Tuberculina maxima* on pycnidia and aecia of *Cronartium ribicola,* the white pine blister rust, in western North America, and found that pycnial fluid is apparently the principal substratum for the growth and development of *Tuberculina.* Pycnia are attacked throughout the period of their production, and a pycnial zone attacked one year usually fails to produce aecia the next season. The aecia which do develop are parasitized and production of aeciospores is reduced. Indications that *T. maxima* also parasitizes the bark well below the rust fructifications on the surface were found when *Tuberculina* sporulated profusely on cut surfaces for at least 1 month after portions of the infected bark of a stem canker were shaved off.[43] Meilke[107] emphasized the lack of information on the nutrient requirements of *Tuberculina,* and also the difficulties of inoculating a rust effectively with its spores for the biological control of *C. ribicola.* Some degree of the control of the rust has been attributed to the mycoparasite, but the ability of the American strain to inactivate cankers has not been encouraging. More virulent strains of *T. maxima* are being sought from those occurring in Europe and Asia. Recently Wicker[174] has reported that aeciospore production has been reduced considerably by *T. maxima.*

Among the hyperparasites of rust fungi, *Cladosporium* has been reported to parasitize a number of rusts.[26,81,105,115,134,136,145] In Australia, in late summer, uredinial pustules of *Melampsora larici-populina* and *M. medusae* are extensively colonized by hyperparasitic fungi, particularly *Cladosporium aecidiicola* which is regarded as a fairly common hyperparasite of rusts in continental Europe and in the mediterranean region.[73,125,128] Sharma and Heather[134,135] found that the presence of conidia of *C. aecidiicola* considerably reduced the viability of urediniospores of *M. larici-populina* and *M. medusae,* and consequently lowered the infection of leaves of *Populus nigra* var. *italica* under controlled conditions. They also demonstrated lysis of germinated and ungerminated urediniospores by the mycelium of *Cladosporium.* Unlike certain hyperparasites, e.g., the *Darluca* sp., there appears to be no close nutritional relationship between *Cladosporium* and rust. While studying the effect of conidia of *C. aecidiicola* on urediniospores of *M. medusae* Sharma and Heather[135] did not observe any mycelial growth in samples containing a mixture of urediniospores and conidia. They concluded that a loss in the viability of urediniospores in mixtures could be due to some volatile inhibitor(s) or leaching of some toxic substance(s) which inactivated the enzyme(s) responsible for mobilizing an endogenous substrate. Working on phylloplane organisms as hyperparasites Omar and Heather[115] reported *Cladosporium* sp., *Alternaria* sp., and *Penicillium* sp. parasitizing urediniospores of *M. larici-populina.* In vivo germination studies of conidia of *Cladosporium* and *Alternaria* caused major reduction in germination while *Penicillium* had only a marginal effect. Lysis of urediniospores was observed only with *Alternaria* and *Cladosporium.* These two fungi also caused a significant reduction in the numbers of uredinia per unit area developed on leaf discs.

Other species of *Cladosporium (C. herbarum* and *C. tenuissimum)* have been recorded recently by Sharma and Heather[130] on poplar rust, *Melampsora larici-populina.* They demonstrated lysis of urediniospores and reduction in infection due to these hyperparasites. Furthermore, they found a highly significant interaction between *Cladosporium* and different races of *M. larici-populina,*[131] thus suggesting a degree of physiologic specialization in these processes. The most effective species was *C. tenuissimum* followed by *C. herbarum* and *C. cladosporioides. C. tenuissimum* also acted as a postpenetration antagonist against *M.*

larici-populina. When the leaf discs were inoculated with the rust 12 and 24 hr before conidial deposition, uredinial density was reduced by 47 to 49 and 31 to 17%, respectively, depending on the race. The reduction in the number of uredinia was more (60%) when conidia of *C. tenuissimum* were deposited immediately after inoculation of the rust. With the exception of *C. cladosporioides,* reduction in uredinia was significantly higher at 20°C than at 12 or 25°C in the case of the other two species of *Cladosporium.*[132]

Tsuneda and Hiratsuka[165] found a frequent occurrence of *C. gallicola* on western gall rust *(Endocronartium harknessi)* of pines in Canada, which reduced the inoculum potential. Recently, Sharma et al.[136] have reported *C. oxysporum* and *Acremonium recifei* on teak rust, *Olivea tectonae. C. oxysporum* was observed just after the rainy season during September to November. The hyperparasite sporulates profusely over the uredinia, often covering a large part of the leaf. *A. recifei* was the main parasite during the comparatively dry period of the year from October to March. Often uredinia were completely covered with the parasitic growth and appeared white. The infected urediniospores were found to be decolorized and lysed. It was suggested that both the hyperparasites play a significant role in checking the epidemic spread of teak rust in Kerala, India.

Apart from the common rust hyperparasites, *Darluca, Tuberculina,* and *Cladosporium,* numerous others have been reported on various rusts. Wollenweber[175] described the observations of L. N. Goodding on the destructive parasitism of *Fusarium bactridioides* on *Cronartium harknessii, C. ribicola,* and *C. filamentosum.* Levine et al.[95] noted a *Trichoderma* sp. which overgrew pustules of stem rust on cereals and infected urediniospores through the germ pores. Ciferri[35] reported that *Monosporium uredinicolum* parasitized *Coleosporium ipomoeae,* the rust of sweet potatoes. In tropical Africa, certain species of *Titaea* (especially *T. hemileae)* have been found to be parasitic in the sori of coffee rust, *Hemileia vastatrix.*[68] On the same rust *Verticillium hemileae* has also been recorded by Bourriquet[29] and Locci et al.[96] The mycelium of the hyperparasite penetrates the urediniospores and causes lysis. *Verticillium lecanii* parasitizes both insects and rusts, including coffee rust, *H. vastatrix.*[87] In temperate regions, *V. lecanii* has been reported on carnation rust *(Uromyces dianthii),*[147] bean stem rust *(Uromyces appendiculatus),*[6,64] and on wheat stem rust *(Puccinia graminis* f. sp. *tritici).*[105] In the presence of *V. lecanii,* infection by *U. dianthii* was prevented or the formation of urediniospores was arrested, depending upon the time of application of conidia of the hyperparasite.[147] The number of erumpent uredinia was also reduced when *V. lecanii* was spread on the plants as the first uredinia had begun to erupt.[148] *V. lecanii* reduced rust on a susceptible cultivar of carnation by ca. 40%, but did not affect the incidence on a resistant one.[149] The hyphae of *V. lecanii* penetrate and invade the urediniospores of bean rust, *U. appendiculatus,* but do not lyse them.[6] In glasshouse trials where bean *(Phaseolus vulgaris)* plants were sprayed with a spore suspension of *V. lecanii* 10 days after the rust infection, spread of the rust to neighboring uninoculated plants was reduced by 68% compared with unsprayed controls. However, attempts at biological control in the field were unsuccessful, probably due to insufficiently high temperature and relative humidity.[64] *Trichothecium roseum* has been reported to parasitize rust of *Saccharum officinarum* and other cereals, and cause degeneration of urediniospores.[3,34] Pruszynska-Gondek[121] found that mycelium of *T. roseum* which grows on the surface of sori of *Uromyces fabae* on broad bean leaves inhibited the urediniospore development.

Tsuneda et al.[166] studied hyperparasitization of *Scytalidium uredinicola* on western gall rust *(Endocronartium harknessii)* of pines, and reported that it destroyed more than 80% of *E. harknessii* galls in some localities in Alberta. Tsuneda and Hiratsuka[165] found parasitization of pine stem rusts, *Cronartium coleosporioides* and *E. harknessii* by *Monocillium nordinii.* Hyphae of *M. nordinii* penetrated the rust spores and many conidiophores, and conidia developed on the host spores. Eventually the spores were completely degraded. They isolated antifungal metabolites from *M. nordinii* which killed the spores. The metabolites include the known compound monorden and five new substances, monocillin I, II, III, IV, and V. Antifungal activity of these metabolites was tested against numerous fungi where morphological abberations, degeneration, and lysis of spores of the test fungi were observed.

b. Bacteria

Fructifications of rusts have also been reported to be parasitized by bacteria. Destruction of sori was brought about by lysis. According to Borders[28] *Erwinia urediniolytica* attacks pedicels of rust spores. Another species, *E. uredovora*, parasitizes the cereal rusts causing degeneration of urediniospores.[69] Pon et al.[119] studied the parasitism of *Xanthomonas uredovorus* on *Puccinia graminis*, which grows on or in urediniospores at a high temperature (30°C) and in a saturated atmosphere. The bacteria which inhabit the soil are spattered by rain up to the rust sori on the lower part of the plant. It was suggested that the temperature requirements of the parasite may however, limit its biological role. A species of *Xanthomonas* has also been observed to affect the viability of urediniospores of *Olivea tectonae* affecting teak during late summer in Kerala, India (J. K. Sharma, unpublished observation). Other bacteria associated with the aerial surface of plants or fungal spores may exhibit a relatively nonspecific action against fungi. For example, three different species of *Bacillus* were shown to lyse urediniospore germ tubes[110] and cultures filtrates of all three species were also able to lyse germ tubes. However, the filtrate from *B. pumilus* could still cause lysis after autoclaving, which indicated that a substance other than an enzyme was involved. French et al.[61] reported that urediniospores of wheat stem rust were always contaminated with various bacteria, especially *Pseudomonas flourescens*, *Bacillus megaterium*, *B. subtilis*, and *B. polymyxa*. Only *P. flourescens* and *B. megaterium* inhibited the urediniospores completely. *B. cereus* was usually found associated with urediniospores of the leek rust, *Puccinia allii*.[49] The bacterium was shown to inhibit germination of urediniospores of the rust in vitro; if applied in sufficiently large concentrations ($1 \times 10^9/m\ell$) to leaves of leek plants a significant reduction in the number of rust pustules was obtained.

2. Leaf Spots

Unlike rusts where the fungus produces external fructifications, reproductive structures such as ascocarps, pycnidia, and sporophores of leaf spot pathogens are not common. The chances of a secondary inoculum being parasitized are considerably limited, except in such cases where the pathogen sporulates on dead infected plant material. Mainly the primary inoculum, which causes the initial disease, may be parasitized thus affecting the incidence of a disease. Hence, hyperparasitism has a limited role to play in the biocontrol of leaf spot diseases. As compared to rusts, only a few examples are available where a leaf spot pathogen is attacked by mycoparasites. In some instances necrotrophic hyperparasitic interactions have been used to control leaf spot diseases.

a. Fungi

Nectria inventa, which parasitizes *Alternaria brassicae*, appears to be stimulated by exudates of its host fungus[167] and its possible use has been suggested for reducing inoculum of the plant pathogen on fallen dead leaves and other plant debris.[168] Kenneth and Issac[83] reported that *Cephalosporium* causes hyperplasia and degeneration of hyphae and spores of *Helminthosporium sativum* and *H. teres*, and consequently suppresses the growth of the pathogens. Recently, *Hansfordia pulvinata* has been recorded to parasitize pathogens of various leaf spot diseases. *H. pulvinata* attacks conidia of *Cladosporium fulvum* on tomato leaves destroying them completely.[117] It was suggested that *H. pulvinata* could be used in the biological control of *C. fulvum* in glasshouse-grown tomatoes. The possible biological control of *Sclerotinia sclerotiorum* on the leaves of bean *(Phaseolus vulgaris)* by a hyperparasite, *Coniothyrium minitans*, was attempted by Trutmann et al.[163] They found that the growth of *C. minitans* was restricted on bean leaves and this was partly associated with the presence of an inhibitor in the leaf surface wax. Spraying trials in the glasshouse and the field established that aerial application of the hyperparasite did not control the disease even though parasitism of mycelium of *S. sclerotiorum* was demonstrated. This was because of inhibition of *C. minitans* on the leaf surface. However, the number of sclerotia produced on *C. minitans* treated plants was reduced to 55% as compared to controls. A species of *Botrytis* has often been found to be associated as a hyperparasite with *Phaeoramularia capsicola* causing velvet spot of pepper *(Capsicum annuum)* in Brazil.[55] It has been suggested that the hyperparasite might be exerting some pressure in the control of the disease.

b. Bacteria

Lenne and Parbery[94] observed clusters of bacteria surrounding lysed conidia and germ tubes of *Colletotrichum gloeosporioides* on leaves. Because bacteria normally failed to lyse appressoria it was suggested that these might survive under conditions where germ tubes and conidia were destroyed.

Lysis of bacteria can be brought about by specialized predatory Bdellovibrios. These become attached to the surface of Gram negative bacteria, penetrate the host cell wall, and enter the space between the cell membrane and wall. Growth and cell division leads to the production of 5 to 6 progenies within about 5 hr and the host cell wall is then lysed, releasing the progeny. The effect of *Bdellovibrio bacteriovorus* on the control of bacterial blight of soybean leaves caused by *Pseudomonas glycinea* has been investigated by Schreff.[129] He concluded that on aerial plant surfaces *B. bacteriovorus* may be useful in reducing disease incidence by bringing about a change in the ecological balance of resident organisms.

B. Antagonism

Naturally occurring microorganisms on aerial plant surfaces, which possess an effective antagonistic action against a pathogen, have been explored for biological control purposes. Organisms adapted to the same habitat as the pathogen are generally preferred over those from other habitats, as the latter are less likely to survive for long in the ecosystem and consequently would have to be reapplied to foliar surfaces more frequently.[24] Porter[120] was possibly one of the first to investigate the interalia interactions of a large number of microorganisms including bacteria, yeasts, and filamentous fungi from a variety of sources. He also examined a number of them for antagonistic activity in vitro against *Helminthosporium sativum* and subsequently used the most promising ones in in vivo experiments against the pathogen on wheat seedlings. He obtained inhibition of growth of the pathogen in culture, and inhibition of the symptom development on seedlings. First experiments on biological control of plant pathogens with antagonists were conducted in Canada nearly 60 years ago by Stanford.[151] Since then a considerable amount of knowledge about the role of antagonists in biological control has been accumulated. Much more research is being done with necrotrophic fungi than with biotrophic fungi. This is possibly because the latter fungi are less susceptible to antagonism during prepenetration development than fungi which are normally nutrient dependent in this stage of development. Mechanism of antagonism has been discussed in detail by Fokkema.[57]

1. Naturally Occurring Antagonists
a. Rusts

There are only a few reports of antagonism against foliar rust fungi.

i. Fungi

Sinha[140] and Kapooria and Sinha[80] demonstrated that addition of spores of several fungi, such as *Chaetomium globosum, Fusarium oxysporum, Trichoderma koningii, Aspergillus japonicus,* and *A. niger* isolated from the leaves of *Pennisetum typhoides,* to an inoculum of *Puccinia penniseti* reduced the number of pustules on the host. All the antagonists inhibited the germination of urediniospores in in vitro studies. Development of *Melampsora occidentalis* on needles of *Pinus trichocarpa* could be reduced by artificial introduction of *Trichoderma* and *Epicoccum.*[21] Similar results were also obtained with wheat rust caused by *Puccinia graminis* f. sp. *tritici.*[108] Considerable reduction in germination of urediniospores in in vitro tests was brought about with culture filtrates of most of the phylloplane fungi. However, in in vivo tests *Penicillium notatum, Myrothecium roridum, Cladosporium herbarum,* and *Nigrospora sphaerica* caused the most inhibition and, consequently, reduction in the number of rust pustules. Recently, Stolbova[155] has shown antibiotic activity of strains of *Cladobotryum varium* against *P. graminis.* Wheat leaves inoculated with *P. graminis* were treated with culture liquids of different strains of *C. varium* which showed varying degree of antirust activity; strain 16 A reduced infection up to 11%. In further tests, two treatments of wheat plants treated with culture liquid of strain 16A decreased the rust development by 35 to 40%. The treatments were given during the appearance of early pustule development and following the second generation of the pathogen. Similarly, Stewart and Hill[153,154] reported effective inhibition of urediniospores of *P. graminis* in vitro and on the leaf surface with culture filtrates of *Helminthosporium sorokinianum,* but a spore suspension proved to be less effective. The mechanism of control of *P. graminis* by *H. sorokinianum* appears to be different to the inhibition reported by Stolbova,[155] Misra and Tiwari,[108] and Kapooria and Sinha.[80]

ii. Bacteria

Recently, the inhibitory effect of an isolate of *Bacillus subtilis* (APPL-I) on the development of rust pustules of bean rust, *Uromyces phaseoli,* has been reported by Baker et al.[10] This isolate gave more than 95% reduction in the subsequent number of rust pustules when it was applied in liquid culture to plants in the glasshouse 2 to 120 hr before inoculation with urediniospores. When APPL-1 was applied after inoculation with the rust there was no effect on the pustule number. Microscopic observation of *B. subtilis* treated leaves showed that urediniospore germination was greatly reduced and no normal germ tubes were produced. Some urediniospores developed abnormal cytoplasmic protrusions. An inhibitory component, which was heat stable and nondialyzable with a molecular weight of 5 to 10 kDaltons, was isolated from the culture filtrate; this component was inhibitory to spore germination and rust severity. In field trials the rust severity was reduced by at least 75% in 1982 and 1983 with three applications per week of *B. subtilis.* In some trials bacterial treatment was more effective than the weekly mancozeb application.

b. Leaf Spots

Interaction between common phylloplane fungi and a variety of necrotrophic leaf spotting pathogens have been demonstrated under more or less controlled environmental conditions by applying the antagonists to the leaves before or at the same time as the pathogen. Information concerning the biological control of leaf spot diseases in the field, however, is very limited. One of the earlier examples of biological control of a disease under field conditions is provided by Bhatt and Vaughan.[19] They obtained a decrease of ca. 40% in *Botrytis* infection of strawberries in the field by spraying with a conidial suspension of *Cladosporium herbarum.* An increase in yield from 6.8 to 8.4 t/acre was obtained when strawberry plants were sprayed three times with *Cladosporium* conidia in 1% glucose, starting at the blossom stage. Fruit rot was not prevented, but the early colonization of flowers and green fruits by *Cladosporium* diminished *Botrytis* attack of fruits. Recently, Islam and Nandi[74,75] have shown control of brown spot of rice *(Drechslera oryzae)* by *Bacillus megaterium.* In field experiments spraying with a cell suspension (1×10^4 bacterial cells/mℓ) at 15 day intervals until grain maturity greatly reduced the disease incidence, and also improved plant growth and crop yield. The antagonist was not phytotoxic to rice; germination of wheat seeds was reduced while that of maize and pea increased.

i. Fungi

There are numerous reports on reduction in infection by antagonists indicating their great potential in biological control. Various studies have shown resident yeasts as promising antagonists. Akai and Kuramoto[4] obtained a 50% reduction of infection of rice leaves with a mixed inoculum of *Cochliobolus miyabeanus* and a *Candida* sp. The antagonist, a common phylloplane yeast, did not, however, inhibit *C. miyabeanus* on agar plates nor did it affect its spore germination. In a similar study Van den Heuvel[170] observed that the number of lesions caused by *Alternaria zinniae* on dwarf bean *(Phaseolus vulgaris)* was greatly reduced by many epiphytic microorganisms such as *Alternaria, Aspergillus, Fusarium, Aureobasidium, Cladosporium, Phoma,* and certain actinomycetes and yeasts. He found that the inhibitory activities of most epiphytic microorganisms in vivo did not agree with their capacities to antagonize the development of *A. zinniae* on agar. In the case of *Alternaria solani,* causing leaf disease in tomatoes, a 2-day preinoculation with *Aureobasidium pullulans* significantly reduced infection and growth of the pathogen;[30] simultaneous inoculation of the antagonist and pathogen on the leaves was ineffective in reducing the disease. On the contrary, when sugar beet leaves were inoculated simultaneously with the parasite, *Phoma betae* and phylloplane microorganisms such as *A. pullulans* or *Torulopsis candida,* the development of expanding lesions was curtailed.[172] The effect was more pronounced when pollen was added on. Fokkema and Van der Meulen[60] reported that when *Aureobasidium pullulans, Sporobolomyces roseus,* and *Cryptococcus laurentii* var. *flavescens* were added to the inoculum, superficial mycelial growth and infection of wheat by *Septoria nodorum* was inhibited by 50% or more. On treating wheat leaves with *Sporobolomyces roseus, Rhodotorula,* and *Sporobolomyces* sp. before inoculation with *Cochliobolus sativus* and *Leptosphaeria nodorum* under controlled conditions, infection of these two pathogens was greatly reduced under controlled conditions.[98] In field trials, the three yeasts reduced the yield in 1980 when fungal leaf spots were almost completely absent. In 1981, *S. roseus* reduced infection by the pathogen and increased yield, but in 1982 it had no effect on disease or yield. Inhibitory effects of various yeasts *(A. pullulans, Sporobolomyces,* and white yeast, especially *Cryptococcus),* isolated from the phyllosphere of rye, against *Drechslera sorokiniana* were studied by Fokkema.[56] He concluded that antagonism was likely to be due to nutrient competition restricting superficial hyphae development and consequently reducing infection.

Trichoderma, which is known to play an important role in the biological control of soilborne diseases, has been recorded to inhibit the leaf pathogens also. Of the 22 microorganisms tested, 4 *Trichoderma* spp. strongly inhibited mycelial growth of the rice pathogen, *Pyricularia oryzae* at 15, 28, and 35°C.[158] Inhibition by *T. pseudokoningii* and *Aspergillus niger* was >60% at pH 4.1 and 6.2, respectively.[159] *Trichoderma harzianum* and an actinomycete were found to be the most effective inhibitors of *Helminthosporium teres* on barley.[5] The possibility of using *T. viride* in biological control of *Stigmina carpophila,* the causal agent of shot-hole of apricot, has been suggested by Payghami.[116]

Inhibition of pathogens is also brought about by *Aspergillus* suggesting their possible use in biological control. Pretreatment of barley seedlings with the antagonist *Aspergillus variecolor* was very effective in lowering the severity of leaf blotch caused by *Alternaria alternata.*[48] Diffusates of the antagonist also caused significant reduction in the severity of the disease. Antagonism of *Aspergillus* to *Helminthosporium spiciferum* on tobacco leaves has been reported by Chauhan and Grover;[33] the degree of interaction differed on two tobacco varieties used. Similarly, common leaf antagonistic saprophytes reduced *Botrytis cinerea* infection of onion,[59] *Colletotrichum graminicola* on corn,[58] *Pseudocercosporella herpotrichoides* on wheat stems[126](the antagonist *Microdochium bolleyi* reduced the disease by 80%), *Phyllactinia dalbergiae* on *Dalbergia sissoo* by *Cladosporium spongiosum*,[99] and *Crinipellis perniciosa* on cocoa by *Cladobotryum amazonense*.[14]

Interactions between pathogens are also not uncommon, but it is unlikely that they can be utilized effectively in biological control. The early infection of leaf blades by *Helminthosporium sativum* was found to inhibit the later development of *Septoria passerinii* in the leaf sheath of barley.[111,173] Johnson and Huffman[78] reported that the incidence of *Puccinia triticina* on a susceptible wheat variety could be reduced by preinoculation of the leaf with urediniospores of crown rust of oat, *Puccinia coronata*. When wheat seedlings were sprayed with a spore suspension of *Drechslera oryzae (Cochliobolus miyabeanus)*, a nonpathogen of wheat, 3 days before inoculation with *Drechslera sorokiniana, (C. sativus)* seedlings developed considerable resistance.[32]

ii. Bacteria

Naturally occurring epiphytic bacteria on leaves are known to give protection against necrotrophic pathogens. Attempts to use bacteria to control fungal, as opposed to bacterial plant pathogens, have been more numerous. During a study of many isolates of bacteria from cucumber leaves, Leben[90] found one isolate to be markedly antagonistic in vitro against the cucumber anthracnose pathogen, *Colletotrichum lagenarium*. When preparations of this bacteria were sprayed onto cucumber leaves in the greenhouse, the incidence of anthracnose was reduced, but the bacterium failed to protect plants in the field.[93] The effect of drying and sunlight were thought to have caused a rapid loss of viability of the applied bacterial cells. An antagonistic bacterium isolated from *Bipolaris maydis* conidia failed to protect corn plants from this pathogen in the field for similar reasons.[143]

Austin et al.[7] showed that a number of phylloplane bacteria, particularly *Pseudomonas fluorescens*, isolated from leaves of *Lolium perenne* were antagonistic to the pathogen *Drechslera dictyoides*. The antagonists reduced spore germination and germ tube growth and caused lysis of hyphae, which led to a reduction in lesion development. *Pseudomonas fluorescens* occurs widely as a phylloplane and rhizosphere inhabitant. Kloepper et al.[86] reported that siderophore producing strains of this bacterium inhibit the growth of other root microorganisms, especially weak pathogens, as a result of complexing available iron. If phylloplane strains possessed similar properties, or if root strains were able to colonize leaf surfaces, a useful biocontrol method for foliar surfaces might result. In contrast, it has been shown that siderophore-producing leaf and fruit surface bacteria can stimulate germination and infection of certain leaf-infecting pathogens.[157] Blakeman and Fraser[25] found inhibition of spore germination of *Botrytis cinerea* in infection droplets. On wetting the leaves bacteria can multiply rapidly from low numbers of surviving organisms,[22] probably existing in protected positions on the leaf.[91]

Purkayastha and Bhattacharyya[122] have reported two strains of *Bacillus megaterium, Aspergillus nidulans,* and *Penicillium oxalicum* isolated from the phyllosphere of jute highly antagonistic to *Colletotrichum corchori*. An ethyl acetate soluble, partially thermolabile, antifungal substance detected in the culture filtrate of *B. megaterium* reduced spore germination of *C. corchori* and caused a >30% reduction in the germ tube length at 0.25 dilution. Foliar sprays of a bacterial suspension and culture filtrates 24 hr prior to inoculation markedly reduced the formation and spread of lesions on jute leaves by *C. corchori;* a bacterial suspension was more effective than its culture filtrate in reducing lesion production.[20] Recently, inhibition of some fungal pathogens in the phylloplane of rice by *B. megaterium* has also been reported.[74,75] *B. megaterium*, which was predominant in the phylloplane, was antagonistic to *Drechslera oryzae (Cochliobolus miyabeanus), Alternaria alternata*, and *Fusarium roseum*.

Application of antagonistic phylloplane bacteria to leaves is probably more successful for control of plant pathogenic bacteria rather than fungi,[16,127] possibly because conditions which favor multiplication of the saprophytes will be very similar to those for the pathogen. Species of antagonistic bacilli, especially *Bacillus cereus* and *B. mycoides,* were found to be part of the natural epiphytic bacterial flora of needles of Douglas fir.[103] When these organisms were applied in nutrient broth to needles, control of *Melampsora medusae* was obtained. A yellow bacterium described by Adam and Pugsley[1] was intimately associated with *Pseudomonas phaseolicola* in all stages of "halo blight" of bean. The bean-saprophytic bacteria was antagonistic and when inoculated simultaneously with *P. phaseolicola,* retarded the development of the disease and diminished its ultimate severity. The yellow bacterium was not specific to bean. An indistinguishable organism isolated from mulberry blight lesions also inhibited infection of bean by *P. phaseolicola* and mulberry by *P. mori.*[1] The role of yellow forms of bacteria in antagonism has been confirmed by Crosse.[39] He found that though nonchromogenic bacteria were also implicated in interactions with bacterial diseases, yellow forms were more frequent.

A number of field experiments using bacteria as biocontrol agents have also given discouraging results because it has been proved that it is not possible to maintain cell numbers at high enough levels to give control, particularly under dry conditions in sunlight.[93,143] In exceptions where the bacterial antagonists were effective, they belonged to genus *Bacillus,* which is capable of forming spores. It is this characteristic which may have enabled their higher population to survive on leaves.

iii. Actinomycetes

Actinomycetes residing on leaves have also been reported to control diseases. On *Brassica campestris,* Sharma and Gupta[139] noted that a decline in the populations of *Streptomyces rochei* was accompanied by a rise in the population of two pathogens, *Alternaria brassicae* and *A. brassicicola,* indicating a possibility of antagonism by the actinomycete. *S. rochei* isolated from the phyllosphere of *Brassica campestris* var. *dichotoma* reduced the conidial germination and infection of *Alternaria brassicae* (60% and 11.3%) and *A. brassicicola* (64.5% and 10.6%), when pathogen and antagonist were inoculated together on seedlings.[137,138] *Streptomyces griseus* on *Capsicum* sp. is shown to antagonize the chilli blight pathogen, *Alternaria solani.*[67] Studies have been suggested for a possible biocontrol of *A. solani* through *S. griseus.*

2. Foreign Antagonists

Antagonistic organisms occurring in other habitats (plant or soil) have also been utilized in the biocontrol of plant diseases. Although it is argued that they may not be able to survive for long in the introduced ecosystem, research done so far indicates their great potential in reducing the disease incidence. Only examples for leaf spot diseases are cited here as none is available for rusts.

a. Fungi

The use of *Trichoderma,* a common soil fungus, has been attempted in the biological control of a number of plant diseases. Biological control of infection of pruning wounds has been found to be successful.[38] The possible explanation given is that growth of *Trichoderma* is favored by the absence of competition and presence of readily available nutrients. Unexpectedly, the *Trichoderma* spp. have also been effective in controlling *Botrytis* infection of undamaged tissues.[52,161] It was possible to control naturally occurring dry-eyespot on windfall apples caused by *B. cinerea* by spraying with *Trichoderma harzianum* in 0.1% malt extract at similar frequencies as currently used fungicides.[162] Equally successful was field control of Botrytis rot of strawberries by spraying with the *Trichoderma* spp. at the time of harvest and during storage.[161] Similarly, Botrytis rot of grapes was controlled in large-scale field trials by *T. viride*.[51,52] Four applications of *T. viride* containing 10[8] conidia/ mℓ, sprayed from the beginning of flowering until 3 weeks before harvest, reduced the infection from 93 to 70% and eventual rot from 32 to 9%. *T. viride* also gave protection against dead arm disease of vine at the dormant bud stage caused by *Phomopsis viticola*.

Rai and Singh[124] studied in detail the antagonistic activity of some leaf surface microfungi against pathogenic fungi, *Alternaria brassicae* and *Drechslera graminea*, in the laboratory and field. The most antagonistic fungi for *A. brassicae* were *Epicoccum purpurescens*, *Aureobasidium pullulans*, and *Cladosporium cladosporioides*, and for *D. graminea*, *Alternaria alternata* and *Aureobasidium pullulans*. Application of spores of the test fungi to leaves, either collectively or individually, inhibited lesion development by the pathogen. Inhibition increased with increasing spore concentration and was highest when a composite spore mixture was used.

b. Bacteria

A bacterial isolate (A180) from cucumber was found to be antagonistic to a number of fungi when tested in dual cultures. Washed bacterial cells also reduced lesions caused by *Alternaria solani* on tomato seedlings and *Trichometasphaeria turcica* on maize seedlings. Bacteria applied with nutrients reduced disease more than cells carried in water. An antifungal antibiotic was extracted from liquid cultures.[90,92] In pot tests *Erwinia herbicola* applied to seed improved emergence and controlled coleoptile streak and leaf stripe symptoms caused by *Pyrenophora avenae* on oat seedlings.[50] The bacterium was transferred from seed to aerial portions of oat, barley, and wheat seedlings. Reduction of incidence of fire blight of pear in an orchard was achieved by repeated spraying during blossom time with an *Erwinia* sp. and three Pseudomonads from infected sugar beet.[160]

Peanut Cercospora leaf spot and tobacco Alternaria leaf spot have been successfully controlled by spraying bacteria from other habitats on the foliage at 7- to 14-day intervals.[150] The organisms were *Pseudomonas cepacia* isolated from conidia of *Bipolaris maydis* obtained from infected corn leaves, *Bacillus mycoides* from tobacco leaves, and *B. thuringiensis* as used in commercial formulation of insect control. Two foliar sprays of a suspension of *Bacillus subtilis* on 30- and 60-day-old rice plants prevented primary and secondary infection of brown rot disease caused by *Bipolaris oryzae* both at seedling and tillering stages.[112]

c. Actinomycetes

There are only two examples available where a foreign antagonistic actinomycete has been used in the control of a disease. *Sorghum* anthracnose caused by *Colletotrichum graminicola* was controlled by spraying pot-grown plants with a suspension of *Streptomyces ganmycicus* at the same time as applying spores of the pathogen.[141] When *S. ganmycicus* was sprayed 1 week prior to the pathogen, disease control was better. The incidence of leaf blotch disease of barley caused by *Alternaria alternata* was reduced considerably when diffusates of *Streptomyces olivaceus* were sprayed on seedlings.[47]

C. Immunization

Immunization is a host mediated response caused by an interruption of the classical disease triangle by potential antagonists.[101] It involves protection against a subsequent attack by a pathogen and it can be invoked by truly nonpathogenic organisms, by a virulent pathogen, or by pathogenic taxa on aberrant hosts. Heat treated spores or sterile culture filtrates may also produce the same response.[101,176] The protective response can usually be brought about by pre-inoculation with the chosen fungus.

In some cases the antagonistic effects of a microorganism were only exerted on the leaf (in vivo) and not in vitro, which suggested the production of inhibitory substances by the leaf under the influence of the antagonist. *Aureobasidium pullulans*, which reduced the number of lesions of *Alternaria zinniae* on bean leaves, did not inhibit spore germination of the parasite in vitro, but spore germination on the leaf was markedly decreased.[170] Similarly, Akai and Kuramoto[4] found that the *Candida* sp., which decreased the amount of brown leaf spot caused by *Cochliobolus miyabeanus* on rice leaves, inhibited neither spore germination nor mycelial growth of the parasite in vitro.

Fungitoxic, phytoalexin-like substances were supposed to inhibit infection when leaves were protected from a disease by an avirulent strain of the pathogen or by a fungus clearly related to the pathogen. The size and number of lesions caused by *Pyricularia oryzae* on rice leaves were reduced by less virulent or avirulent strains of the pathogen. This reduction was associated with the formation of two phytoalexin-like compounds in and around the hypersensitive lesions induced by an invasion of the avirulent race.[8,85,114] Similar observations have been made for *Cochliobolus miyabeanus* on *O. sativa*,[142] *Colletotrichum orbiculare* on *Citrullus vulgaris*,[106] and *Colletotrichum lindemuthianum* on *Vicia faba* and *Phaseolus vulgaris*.[44] In these cases protection from the pathogen is possibly due to the formation of high levels of an inhibitor, a phytoalexin. Rahe et al.[123] obtained a hypersensitive response in hypocotyls of *P. vulgaris* by inoculating them with *Helminthosporium carbonum*, an *Alternaria* sp., or avirulent *C. lindemuthianum*. The inoculated tissues were subsequently resistant to pathogenic races of *C. lindemuthianum*. A different explanation for the immunization effect is given by Johnson and Huffman[78] who inoculated *Triticum* seedlings with *Puccinia coronata* obtained from *Avena sativa*. Pustule formation by *Puccinia recondita* subsequently inoculated onto these seedlings was inhibited. They suggested that the previously inoculated rust invaded the substomatal chambers and obstructed the stomatal opening, preventing infection via this route.

D. Hypovirulence

Hypovirulence, which literally means subnormal virulence, could be a consequence of hyperparasitism and hybridization between the alien pathogen and less virulent native relatives.[53] Low virulent variants of plant pathogens have long been known but this loss of virulence by pathogens has not been explored for its potential in disease control.[169] There are no reports available on this aspect either for rusts or leaf spot diseases. The use of hypovirulent strains to balance the pathosystem has interesting possibilities in the biological control of aerial diseases of plants. Hypovirulence, a phenomenon first associated with the chestnut blight fungus, is briefly dealt with here to signify its immense potential in the biocontrol of pathogens of aerial plant parts.

Distinction between types of hypovirulence are based primarily on whether the genetic information that determines them is nuclear or cytoplasmic. Nuclear hypovirulence is due to nuclear genetic determinants and might be caused by mutant nuclear genes, heterokaryons, hybrid nuclei, or extrachromosomal genetic determinants that reside in the nuclei. Cytoplasmic hypovirulence (CH) is through agents such as a virus, virus-like agents, and plasmids and organelles that carry genetic elements, such as mitochondria.[53] According to Elliston[53] it is likely that genetic determinants conferring hypovirulence may reside simultaneously in the nuclei and cytoplasm of a pathogen, and also at times nuclear and cytoplasmic determinants operate together to reduce the pathogenicity of an organism.

The popular interest in transmissible hypovirulence (TH) stems from the findings of Grente and Sauret[66] that certain isolates of *Endothia parasitica* obtained from abnormal cankers of chestnut trees in Italy, retarded or prevented canker development when inoculated into chestnut bark together with related normal isolates. Berthelay-Sauret[17] found that the factors conferring exclusive hypovirulence in *E. parasitica* were cytoplasmic. Subsequently, Grente and Berthelay-Sauret[65] developed a biocontrol program of chestnut canker in France using CH strains. Day et al.[41] found that CH strains of *E. parasitica* consistently contain double stranded ribonucleic acid (ds RNA), genetic material typical of most fungal viruses, and so it is reasonable to assume that virus-like agents are responsible for this condition.[53] Application of hypovirulent strains in the control of chestnut canker in the U.S. was initially confronted with incompatibility, preventing the transfer of ds RNA. This was overcome by applying mixtures of hypovirulent strains. Since a treatment of individual cankers in the forest is impossible this method of disease control is practical only if the applied inoculum is sufficiently pathogenic to sustain itself in the forest.[77]

The Dutch elm pathosystem involving *Ceratocystis ulmi* and the *Ulmus* spp. is another example of hypovirulence. Though ds RNA is reported in the aggressive strain of *C. ulmi*, the role of the cytoplasmic agents in determining the degree of aggressiveness of the strains is not understood.[53] Cytoplasmic hypovirulence has also been described in an isolate of *Rhizoctonia solani* by Castenho et al.[31] The isolate reduced incidence of damping-off in sugar beet seedlings when incorporated into the soil with the seed. However, the effect was short-lived.

Based on the same prinicple of hypovirulence, biological control of the crown gall pathogen, *Agrobacterium radiobactor* var. *tumefaciens* by closely related, nonpathogenic *A. radiobactor* var. *radiobactor* has been achieved. Working on the mechanism of the biological control of the crown gall pathogen Kerr and Htay[84] found that the nonpathogen produces a bacteriocin which inhibited pathogenic strains. They concluded that long-term control of crown gall could result from establishment of a suitable nonpathogenic strain around the roots and crown of the plants.

It is possible that strains with CH agents are a more common occurrence in fungi than known. Attempts are, however, necessary to detect them and to tap their potentials in the biocontrol of foliar diseases.

III. FUTURE PROSPECTS

In assessing the potential for biological control of diseases in agricultural and forestry crops it is at once fairly obvious that the prospects are more promising in agriculture. The annual habit of most agricultural crop plants and the corresponding short rotational interval between successive crops provide greater opportunity for cultural manipulation compared to that of the long-term rotation of forest-tree stands. The high capital investment in land and machinery, and the high economic return of an agricultural crop, justify expenditure on intensive cultural practices and might reasonably be expected to provide greater possibilities for biological control. These considerations would apply in fruit tree orchards and in forest nursery operations. Biological control in practice, apart from special situations, is therefore often possible only within the framework of integrated control, which itself normally depends upon a core of biological control and plant resistance. Moreover, the strategy of integrated control is to reduce the disease incidence and severity within the economical limits, not prevention, which is truly consistent with the biological control. A system of integrated control is dynamic and must be considered as a unit in respect to application of ecological principles used in various methods employed for disease suppression.

Biological control on aerial plant surfaces is much less well developed than in the soil/rhizosphere environment. For this there are two main reasons: (1) availability of cheap and effective fungicides and their ease of application to plant foliage, and (2) antagonists can maintain themselves more readily in the soil because of the more uniform environment. These have discouraged the use of biological control against foliar diseases. Many factors have to be considered in deciding whether a biological system is feasible for control of a particular pathogen. Of prime importance is the availability of a suitable microorganism, a hyperparasite, or an antagonist capable of maintaining itself on the host plant. Therefore, it is just not enough to show growth inhibition of a pathogen or reduction in severity of a disease by an organism. A similar emphasis needs to be placed on their persistence in the ecosystem. The environment under which the crop is grown will play a significant part in determining whether effective population levels of an antagonist/hyperparasite can be established in competition with the existing microflora. The environment may also govern the choice of an organism. For example, yeast and spore-forming bacteria can survive on leaves more readily than nonspore-forming bacteria under adverse conditions.[93,143] It is essential that the primary mechanism by which antagonism is brought about should be known. A biotrophic pathogen may be inhibited by a quite different mechanism from a necrotrophic pathogen. Therefore, for an antagonist to be used as a biocontrol agent it should have an extremely high reproductive capacity, persistence under unfavorable environmental and nutritional conditions, and be extremely aggressive or antagonistic. Unless an antagonist possesses these attributes, it would have little advantage over conventional protective fungicides. The hyperparasites, though they may be highly efficient, have a limited role to play in the control of plant diseases as compared to antagonists. Diseases caused by biotrophic pathogens such as rusts and mildews, which sporulate profusely on the host, may be controlled more effectively by hyperparasites. However, Kranz[87] states that hyperparasites are unlikely to be able to control rusts and mildews alone, but they may be "valuable components in an integrated control system." This is perhaps true for naturally occurring or introduced microorganisms in relation to the control of both biotrophic and necrotrophic pathogens. Although hyperparasites may play a regulatory role in epidemics of such diseases, it remains questionable whether control will be sufficient in agricultural crops, where even small losses are unacceptable. On the other hand a moderate pathogen population is mostly a prerequisite for the survival of the hyperparasite. The most ideal hyperparasite suitable for biological control should: (1) grow rapidly and sporulate profusely under a variety of environmental conditions, (2) infect a parasite throughout its ecological range, (3) attack early, before substantial damage is done to the host by the parasite, (4) have the capacity to debilitate or kill the parasite, (5) have the capability to persist under adverse environmental conditions, and (6) be nonpathogenic to other higher plants.

Despite the experimental evidences presented by examples of rust and leaf spot diseases, no widely accepted application in biocontrol has emerged. Research on microbial biological control of plant diseases has gained a momentum during the last 15 years, and the importance of epiphytic microflora in disease control in nature does not remain now a matter of conjecture. The review clearly indicates that microbial control has considerable potential for expanded application against plant pathogens. Barriers of experiments confined to a laboratory are being crossed, and more and more field trials are being conducted to test the efficacy of a microorganism in the biocontrol of a disease. This is an important step towards the practical application of results obtained in laboratory or glasshouses. Some of the excellent examples of biocontrol of rust and leaf spot diseases under field conditions are provided by Baker et al.[9,10] *(Uromyces phaseoli* on bean by *Bacillus subtilis),* Islam and Nandi[74,75] *(Drechslera oryzae* on rice by *Bacillus megaterium),* and Thompson et al.[160] *(Erwinia amylovora* on pear by *Erwinia* sp. and *Pseudomonas* spp.). Though these examples clearly establish the role of microorganisms in the control of plant diseases, exhaustive field trials will be required before such results can be put to practice in the field.

There is a strong evidence that natural biological control through epiphytic microorganisms provides protection against many foliar diseases such as rusts and leaf spots in the field. It is conceivable that the natural epiphytic microflora could be enhanced and manipulated, possibly by the use of nutrient sprays. It is important that the indiscriminate use of fungicides may be avoided so that the population of biocontrol agents is not reduced in the habitat.[70] The differential effect of some fungicides on components of leaf surface microflora has been discussed by Fokkema and de Nooij.[58] Broad spectrum fungicides such as Captan® and dithiocarbamates strongly inhibit growth of saprophytic fungi, whereas bacteria remain unaffected. Benzimidazoles inhibit *Sporobolomyces, Cladosporium* spp., *Aureobasidium pullulans,* and to a lesser extent, *Cryptococcus.* In formulating an integrated disease control program, emphasis should also be given on the spectrum of activity of the chemicals to be used. The greater the specificity of action of a chemical against the pathogen which it is desired to control, the more effective the integrated control system, because the unaffected saprophytic microflora can antagonize other potential pathogens.

The advantages of biological control are numerous. In the future, the successful biological control of rusts and leaf spot diseases will depend not only on effectiveness, but also on a competitive cost with conventional fungicides and a lack of side effects of the applied organisms. In contrast to chemical control, other benefits of biological control may include the lessening of long-term damage to the environment by the use of persistent chemicals, and an absence of chemical residues in the environment. But to make large-scale use of biological control of a plant pathogen a reality in the future, greatly expanded research with proper field trials over a long period of time is warranted.

ACKNOWLEDGMENTS

The authors are thankful to Dr. C. T. S. Nair, Director, Kerala Forest Research Institute for encouragement and Mrs. K. Annapoorni, Stenographer, and Mr. E. P. Somasekharan Nair for help in word processing.

REFERENCES

1. **Adam, D. B. and Pugsley, A. T.,** A yellow bacterium associated with "halo blight" of beans, *Aust. J. Exp. Biol. Med. Sci.,* 13, 157, 1935.
2. **Adams, J. F.,** *Darluca* on *Peridermium peckii, Mycologia,* 12, 309, 1920.
3. **Ahmad, S. T.,** *Trichothecium roseum* a hyperparasite of rusts, *Indian Phytopathol.,* 23, 634, 1970.
4. **Akai, S. and Kuramoto, T.,** Micro-organisms existing on leaves of rice plants and the occurrence of brown leaf spot, *Ann. Phytopathol. Soc. Japan,* 34, 313, 1968.
5. **Al-Ali, B., Barrault, G., and Albertini, L.,** *In vitro* action of fungal and bacterial antagonists on the mycelial growth of *Helminthosporium teres* Sacc., barley parasite, *Bull. Trimest. Soc. Mycol. Fr.,* 95, 279, 1979.
6. **Allen, D. J.,** *Verticillium lecanii* on the bean rust fungus, *Uromyces appendiculatus, Trans. Br. Mycol. Soc.,* 79, 362, 1982.
7. **Austin, B., Dickinson, C. H., and Goodfellow, M.,** Antagonistic interactions of phylloplane bacteria with *Drechslera dictyoides* (Drechsler) Shoemaker, *Can. J. Microbiol.,* 23, 710, 1977.
8. **Bailey, J.,** Production of antifungal compounds in cowpea (*Vigna sinensis*) and pea (*Pisum sativum*) after virus infection, *J. Gen. Microbiol.,* 75, 119, 1971.
9. **Baker, C. J., Stavely, J. R., Thomas, C. A., Sesser, M., and Macfull, J. S.,** Inhibitory effect of *Bacillus subtilis* on *Uromyces phaseoli* and on development of rust pustules on bean leaves, *Phytopathology,* 73, 1148, 1983.
10. **Baker, C. J., Staveley, J. R., and Mock, N.,** Biocontrol of bean rust by *Bacillus subtilis* under field conditions, *Plant Dis.,* 69, 770, 1985.
11. **Baker, K. F. and Cook, R. J.,** *Biological Control of Plant Pathogens,* Freeman, San Francisco, 1974, 433.
12. **Baker, R.,** Mechanism of biological control of soil-borne pathogens, *Annu. Rev. Phytopathol.,* 6, 263, 1968.
13. **Barnett, H. L.,** The nature of mycoparasitism by fungi, *Annu. Rev. Microbiol.,* 17, 1, 1963.
14. **Bastos, C. N.,** Effect of the culture filtrate of *Cladobotryum amazonense* on *Crinipellis perniciosa* (Stahel) Singer and other pathogens, *Rev. Theobrom.,* 14, 263, 1984.
15. **Bean, G. A.,** Growth of the hyperparasite *Darluca filum* on chemically defined medium, *Phytopathology,* 58, 252, 1968.
16. **Beer, S. V. and Rendel, J. R.,** Inhibition of *Erwinia amylovora* by bacteriocin like substances, *Phytopathology,* 70 (Abstr.), 459, 1980.

17. **Berthelay-Sauret, S.,** Utilisation de mutants auxotrophes dans les recherches sur le determinisme de ''l' hyovirulence exclusive'', *Ann. Phytopathol.,* 5, 318, 1973.

18. **Bhargava, S. N., Shukla, D. N., and Sing, N.,** Observations on *Darluca* — a hyperparasite, *Proc. Nat. Acad. Sci. India Sect.(B),* 52, 78, 1982.

19. **Bhatt, D. D. and Vaughan, E. K.,** Preliminary investigations on biological control of grey mould (*Botrytis cinerea*) of strawberries, *Plant Dis. Rep.,* 46, 342, 1962.

20. **Bhattacharyya, B. and Purkayastha, R. P.,** Biological control of anthracnose disease of jute, *Curr. Sci. Bangalore,* 51, 429, 1982.

21. **Bier, J. E.,** Some effects of foilage saprophytes in the control of *Melampsora* leaf rust on black cottonwood, *For. Chron.,* 41, 306, 1965.

22. **Blakeman, J. P.,** Effect of plant age on inhibition of *Botrytis cinerea* spores by bacteria on beetroot leaves, *Physiol. Plant Pathol.,* 2, 143, 1972.

23. **Blakeman, J. P., Ed.,** *Microbial Ecology of the Phylloplane,* Academic Press, London, 1981, 502.

24. **Blakeman, J. P. and Fokkema, N. J.,** Potential for biological control of plant diseases on the phylloplane, *Annu. Rev. Phytopathol.,* 20, 167, 1982.

25. **Blakeman, J. P. and Fraser, A. K.,** Inhibition of *Botrytis cinerea* spores by bacteria on the surface of chrysanthemum leaves, *Physiol. Plant Pathol.,* 1, 45, 1971.

26. **Bolland, L.,** Poplar rust in Queensland, *Aust. Plant Pathol. Soc. Newsl.,* 2, 28, 1973.

27. **Boosalis, M. G.,** Hyperparasitism, *Annu. Rev. Phytopathol.,* 2, 363, 1964.

28. **Borders, H. I.,** Unpublished M.S. thesis, University of Minnesota, 1938, cited in, Sinha, S., Microbiological complex of the phyllosphere and disease control. Presidential Address, *Indian Phytopathol.,* 18, 1, 1965.

29. **Bourriquet, G.,** Les maladies des plantes cultiv'ees a Madagascar, *Encycl. Mycol.,* 12, 137, 1946.

30. **Brame, C. and Flood, F.,** Antagonism of *Aureobasidium pullulans* towards *Alternaria solani, Trans. Br. Mycol. Soc.,* 81, 621, 1983.

31. **Castenho, B., Butler, E. E., and Shepherd, R. J.,** The association of double-stranded RNA with *Rhizoctonia* decline, *Phytopathology,* 68, 1515, 1978.

32. **Chakraborty, D. and Sinha, A. K.,** Similarity between the chemically and biologically induced resistance in wheat seedlings to *Drechslera sorokiniana, Z. Pflanzenkr. Pflanzenschutz,* 91, 59, 1984.

33. **Chauhan, M. S. and Grover, R. K.,** Myco-phylloflora of tobacco and their antagonism against *Helminthosporium spiciferum, Indian J. Mycol. Plant Pathol.,* 3, 169, 1973.

34. **Chona, B. L., Durgapal, J. C., Guguani, M. C., and Sohi, H. S. A.,** *Trichothecium roseum* Link a hyperparasite of sugarcane rust (*Puccinia erianthi*), *Indian Phytopathol.,* 18, 386, 1965.

35. **Ciferri, R.,** *Sydowia,* 8, 245, 1954, cited in, De Vay, J. E., Mutual relationships in fungi, *Annu. Rev. Microbiol.,* 10, 115, 1956.

36. **Cook, R. J.,** Biological control of plant pathogens: theory to application, *Phytopathology,* 75, 25, 1985.

37. **Cook, R. J. and Baker, K. F.,** *The Nature and Practice of Biological Control of Plant Pathogens,* American Phytopathological Society, St. Paul, Minn., 1983, 539.

38. **Corke, A. T. K.,** Biological control of tree diseases, *Long Ashton Res. Stn. Univ.,* Bristol Rep., 190, 1980.

39. **Crosse, J. E.,** Interactions between saprophytic and pathogenic bacteria in plant disease, in *Ecology of Leaf Surface Microorganisms,* Preece, T. F. and Dickinson, C. H., Eds., Academic Press, London, 1971, 283.

40. **Darpoux, H.,** Biological interference with epidemics, in *Plant Pathology — An Advanced Treatise,* Vol. 3, Horsfall, J. G. and Diamond, A. E., Eds., Academic Press, New York, 1960, 521.

41. **Day, P. R., Dodds, J. A., Elliston, J. E., Jaynes, R. A., and Amagnostakis, S. L.,** Double-stranded RNA in *Endothia parasitica, Phytopathology,* 67, 1393, 1977.

42. **Debach, A., Ed.,** *Biological Control of Insect Pests and Weeds,* Chapman and Hall, London, 1964, 844.

43. **De Vay, J. E.,** Mutual relationships in fungi, *Annu. Rev. Microbiol.,* 10, 115, 1956.

44. **Deverall, B. J., Smith, I. M., and Makris, S.,** Disease resistance in *Vicia faba* and *Phaseolus vulgaris, Neth. J. Plant Pathol.,* 74 (Suppl. 1), 737, 1968.

45. **Dickinson, C. H.,** Interactions of fungicides and leaf saprophytes, *Pestic. Sci.,* 4, 563, 1973.

46. **Dickinson, C. H. and Preece, T. F., Eds.,** *Microbiology of Aerial Plant Surfaces,* Academic Press, London, 1976, 669.

47. **Dixit, R. B. and Gupta, J. S.,** Studies on the biological control of leaf blotch disease of barley by *Streptomyces olivaceus, Acta Bot. Indica,* 8, 190, 1980.

48. **Dixit, R. B. and Gupta, J. S.,** A possibility of biological control of *Alternaria* leaf blotch disease of barley with *Aspergillus variecolor, Indian J. Mycol. Plant Pathol.,* 12, 10, 1982.

49. **Doherty, M. A. and Preece, T. F.,** *Bacillus cereus* prevents germination of uredospores of *Puccinia allii* and the development of rust disease of leek, *Allium porrum,* in controlled environment, *Physiol. Plant Pathol.,* 12, 123, 1978.

50. **Downes, M. J.,** *Erwinia herbicola* for the control of *Pyrenophora avenae* on oat seedlings and of other organisms, in *Proceedings of Seminar on Biological Control,* Duggan, J. J., Ed., Irish National Committee for Biology, Dublin, 1978, 97.

51. **Dubos, B. and Bulit, J.,** Filamentous fungi as biocontrol agents on aerial plant surfaces, in *Microbial Ecology of the Phylloplane,* Blakeman, J. P., Ed., Academic Press, London, 1981, 353.

52. **Dubos, B., Bulit, J., Bugaret, Y., and Verdu, D.,** Possibilities d'utilisation de *Trichoderma viride* Pers. Commemoyen biologique de lutte contre la pourriture grise (*Botrytis cinerea* Pers.) et l'excoriosa (*Phomopsis viticola* Sacc.) de la vigue, *C. R. Acad. Agric. Fr.,* 64, 1159, 1978.

53. **Elliston, J. E.,** Hypovirulence, in *Advances in Plant Pathology,* Vol. 1, Ingram, D. S. and Williams, P. H., Eds., Academic Press, London, 1982, 1.

54. **Fedorinchik, N. S.,** *Mikrobiologya,* 21, 711, 1952, cited in De Vay, J. E., Mutual relationships in fungi, *Annu. Rev. Microbiol.,* 10, 115, 1956.

55. **Figueiredo, M. B., Pimental, C. P. V., Campacei, C. A., Lasca, C. De. C., and Chiba, S.,** Velvet spot of pepper (*Capsicum annuum* L.) caused by *Phaeoramularia capsicicola* (Vassilijevskiy) Deighton and its control, *Biologico,* 49, 45, 1983.

56. **Fokkema, N. J.,** The role of saprophytic fungi in antagonism against *Drechslera sorokiniana* (*Helminthosporium sativum*) on agar plates and on rye leaves with pollen. *Physiol. Plant Pathol.,* 3, 195, 1973.

57. **Fokkema, N. J.,** Antagonism between fungal saprophytes and pathogens on aerial plant surfaces, in *Microbiology of Aerial Plant Surfaces,* Dickinson, C. H. and Preece, T. F., Eds., Academic Press, London, 1976, 487.

58. **Fokkema, N. J. and de Nooij, M. P.,** The effect of fungicides on the microbial balance in the phyllosphere, *EPPO Bull.,* 11, 303, 1981.

59. **Fokkema, N. J. and Lorbeer, J. W.,** Interactions between *Alternaria porri* and the saprophytic mycoflora of onion leaves, *Phytopathology,* 64, 1128, 1974.

60. **Fokkema, N. J. and Van der Meulen, F.,** Antagonism of yeast like phyllosphere fungi against *Septoria nodorum* on wheat leaves, *Neth. J. Plant Pathol.,* 82, 13, 1976.

61. **French, R. C., Novotny, J. F., and Searles, R. B.,** Properties of bacteria isolated from wheat stem rust spores, *Phytopathology,* 54, 970, 1964.

62. **Garrett, S. D.,** Towards biological control of soil-borne plant pathogens, in *Ecology of Soil-borne Plant Pathogens,* Baker, K. F. and Snyder, W. C., Eds., University of California Press, Berkeley, Calif., 1965, 4.

63. **Gonzalez Avila, M. and Castellanos, J. J.,** Presence of the mycoparasite *Darluca filum* on uredosori of *Uromyces phaseoli* var. *typica, Cienc. Agric.,* 3, 119, 1978.

64. **Grabski, G. C. and Mendgen, K.,** Use of *Verticillium lecanii* as a biological control agent against bean rust fungus *Uromyces appendiculatus* var. *appendiculatus* in the field and in the glass house, *Phytopathol. Z.,* 113, 243, 1985.

65. **Grente, J. and Berthelay-Sauret, S.,** Biological control of chestnut blight in France, in *Proceedings of American Chestnut Symposium 1978,* Mac Dondald, W., Chech, F. C., Luchok, J., and Smith, C., Eds., West Virginia University Books, Morgantown, West Virginia, 1979, 30.

66. **Grente, J. and Sauret, S.,** L'hypovirulence exclusive, phenomene original en pathologie vegetale, *C. R. Acad. Sci. Ser. D.,* 268, 2347, 1969.

67. **Gupta, J. S., Agarwal, S. P., and Sharma, S. K.,** Interaction between the surface microbes of chilli and its leaf blight pathogen *Alternaria solani, Indian J. Mycol. Plant Pathol.,* 12, 92, 1982.

68. **Hansford, C. G.,** The foliicolous Ascomycetes, their parasites and associated fungi, *Mycol. Papers, Imp. Mycol. Inst.,* 15, 1, 1946.

69. **Hevesi, M. and Marshal, S. F.,** Contributions to the mechanism of infection of *Erwinia uredovora,* a parasite of rust fungi, *Acta Phytopathol. Acad. Sci. Hung.,* 10, 275, 1975.

70. **Hislop, E. C.,** Some effects of fungicides and other agrochemicals on the microbiology of the aerial surfaces of plants, in *Microbiology of Aerial Plant Surfaces,* Dickinson, C. H. and Preece, T. F., Eds., Academic Press, London, 1976, 41.

71. **Hislop, E. C. and Cox, T. W.,** Effects of captan on the non-parasitic flora of apple leaves, *Trans. Br. Mycol. Soc.,* 52, 223, 1969.

72. **Huffaker, C. B., Ed.,** *Biological Control,* Plenum Press, New York, 1971, 511.

73. **Hulea, A.,** Contributions a la connaissance des champignoss commensauix des Uredinees, *Acad. Rounaine Sec. Stiintif. Bull.,* 22, 1, 1939.

74. **Islam, K. Z. and Nandi, B.,** Inhibition of some fungal pathogens of host phylloplane by *Bacillus megaterium, Z. Pflanzenkr. Pflanzenschutz,* 92, 233, 1985.

75. **Islam, K. Z. and Nandi, B.,** Control of brown spot of rice by *Bacillus megaterium, Z. Pflanzenkr. Pflanzenschutz,* 92, 241, 1985.

76. **Ito, K. and Chiba, O.,** Biological control of needle cast of conifers, in *Biological Control of Forest Diseases,* Nordin, V. J., Ed., Canadian Forestry Service, Ontario, 1971, 99.

77. **Jaybes, R. A. and Elliston, J. E.**, Pathogenicity and canker control by mixtures of hypovirulent strains of *Endothia parasitica* in American chestnut, *Phytopathology*, 70, 453, 1980.
78. **Johnson, C. O. and Huffman, M. D.**, Evidence of local antagonism between two cereal rust fungi, *Phytopathology*, 48, 69, 1958.
79. **Joshi, L. M.**, Recent contributions towards epidemiology of wheat rust in India, *Indian Phytopathol.*, 29, 1, 1976.
80. **Kapooria, R. G. and Sinha, S.**, Phylloplane mycoflora of pearl millet and its influence on the development of *Puccinia penniseti*, *Trans. Br. Mycol. Soc.*, 53, 153, 1969.
81. **Keener, J. G.**, *Cladosporium aecidiicola* Thum. and *Tuberculina persicinia* (Ditm.) Sacc. associated with *Puccinia conspicua* (Arth.) Mains on *Helinium hoopesii* A. Gray in Arizona, *Plant Dis. Rep.*, 38, 690, 1954.
82. **Keener, P. G.**, Biological specialization in *Darluca filum*, *Bull. Torrey Bot. Club*, 61, 475, 1934.
83. **Kenneth, R. and Issac, P. K.**, *Cephalosporium* species parasitic on *Helminthosporium*, *Can. J. Plant Sci.*, 44, 182, 1963.
84. **Kerr, A. and Htay, K.**, Biological control of crown gall through bacteriocin production, *Physiol. Plant Pathol.*, 4, 37, 1974.
85. **Kiyosawa, S. and Fujiimaki, H.**, Studies on mixture inoculation of *Pyricularia oryzae* on rice. Effects of mixture inoculation and concentration on the formation of susceptible lesions in the injection inoculation, *Bull. Nat. Inst. Agric. Sci. Tokyo Ser.*, 17, 1, 1967.
86. **Kloepper, J. W., Leong, J., Teintze, M., and Schroth, M. N.**, Enhanced plant growth by siderophores produced by plant growth-promoting rhizobacteria, *Nature*, 286, 885, 1980.
87. **Kranz, J.**, Hyperparasitism of biotrophic fungi, in *Microbial Ecology of the Phylloplane*, Blakeman, J. P., Ed., Academic Press, London, 1981, 327.
88. **Kuhlman, E. G.**, Hypovirulence and hyperparasitism, in *Plant Disease: An Advanced Treatise*, Horsfall, J. G. and Cowling, E. B., Eds., Academic Press, London, 1980, 363.
89. **Kuhlman, E. G., Carmichael, J. W., and Miller, T.**, *Scytalidium uredinicola*, a new mycoparasite of *Cronartiuum fusiforme* on *Pinus*, *Mycologia*, 68, 1188, 1978.
90. **Leben, C.**, Influence of bacteria isolated from healthy cucumber leaves on two leaf diseases of cucumber, *Phytopathology*, 54, 405, 1964.
91. **Leben, C.**, Survival of plant pathogenic bacteria, *Ohio Agric. Res. Dev. Cent., Spec. Circ.*, 100, 1, 1974.
92. **Leben, C. and Daft, G. C.**, Influence of an epiphytic bacterium on cucumber anthracnose, early blight of tomato, and northern leaf blight of corn, *Phytopathology*, 55, 760, 1965.
93. **Leben, C., Daft, G. C., Wilson, J. D., and Winter, H. F.**, Field tests for disease control by an epiphytic bacterium, *Phytopathology*, 55, 1375, 1965.
94. **Lenne, J. M. and Parbery, D. G.**, Phyllosphere antagonists and appressorium formation in *Colletotrichum gloeosporioides*, *Trans. Br. Mycol. Soc.*, 66, 334, 1976.
95. **Levine, M. N., Bamberg, R. H., and Atkinson, R. R.**, Microorganisms antibiotic or pathogenic to cereal rusts, *Phytopathology*, 26, 99, 1936.
96. **Locci, R., Ferranti, G. M., and Rodrigues, C. J.**, Studies by transmission and scanning electron microscopy on the *Hemileia vastatrix — Verticillium hemileiae* association, *Riv. Patol. Vegetale*, 7, 127, 1971.
97. **Lupton, F. G. H.**, Biological control: the plant breeders objective, *Ann. Appl. Biol.*, 104, 1, 1984.
98. **Luz, W. C. DA.**, Effect of phylloplane microorganisms on fungal leaf spots of wheat, *Fitopathol. Brasileira*, 10, 79, 1985.
99. **Mathur, M. and Mukerji, K. G.**, Antagonistic behavior of *Cladosporium spongiosum* against *Phyllactinia dalbergiae* on *Dalbergia sissoo*, *Angew. Bot.*, 95, 75, 1981.
100. **Madelin, M. F.**, Fungi parasitic on other fungi and lichens, in *The Fungi: An Advanced Treatise: Vol III, The Fungal Population*, Ainsworth, G. C. and Sussman, A. S., Eds., Academic Press, New York, 1968, 253.
101. **Matta, A.**, Microbial penetration and immunisation of uncongenial host plants, *Annu. Rev. Phytopathol.*, 9, 387, 1971.
102. **McAlpine, D.**, *The Rusts of Australia*, R. S. Brain, Government Printer, Melbourne, Australia, 1906, 349.
103. **McBride, R. P.**, A microbiological control of *Melampsora medusae*, *Can. J. Bot.*, 47, 711, 1969.
104. **McBridge, R. P.**, Micro-organism interactions in the phyllosphere of larch, in *Ecology of Leaf Surface Microorganisms*, Preece, T. F. and Dickinson, C. H., Eds., Academic Press, London, 1971, 545.
105. **McKenzie, E. H. C. and Hudson, H. J.**, Mycoflora of rust-infected and non-infected plant material during decay, *Trans. Br. Mycol. Soc.*, 66, 223, 1976.
106. **McLean, D. M.**, Interaction of race I and race II of *Colletotrichum orbiculare* on watermelon, *Plant Dis. Rep.*, 51, 885, 1967.
107. **Meilke, J. L.**, *Phytopathology* 23, 299, 1933, cited in, De Vay, J. E., Mutual relationships in fungi, *Annu. Rev. Microbiol.*, 10, 115, 1956.

108. **Mishra, R. R. and Tiwari, R. P.,** Studies on biological control of *Puccinia graminis tritici,* in *Microbiology of Aerial Plant Surfaces,* Dickinson, C. H. and Preece, T. F., Eds., Academic Press, London, 1976, 559.

109. **Morelet, M. and Pinon, J.,** *Darluca filum* hyperparasite of the genus *Melampsora* on poplar and willow, *Rev. For. Fr.,* 25, 378, 1973.

110. **Morgan, F. L.,** Infection, inhibition and germ-tube lysis of three cereal rusts by *Bacillus pumilus,* *Phytopathology,* 53, 1346, 1963.

111. **Morton, D. J. and Peterson, G. A.,** An inverse relationship between the severities of *Helminthosporium* leaf-blade and *Septoria* leaf-sheath symptoms on barley, *Plant Dis. Rep.,* 44, 23, 1960.

112. **Nanda, H. P. and Gangopadhyay, S.,** Control of rice helminthosporiose with *Bacillus subtilis* antagonistic towards *Bipolaris oryzae, Int. J. Trop. Plant Dis.,* 1, 25, 1983.

113. **Nordin, V. J., Ed.,** *Biological Control of Forest Diseases,* Candian Forestry Service, Ontario, 1971, 106.

114. **Ohata, K. and Kozaka, T.,** Interaction between two races of *Pyricularia oryzae* in lesion-formation in rice plants and accumulation of fluorescent compounds associated with infection (In Japanese, English Summary), *Bull. Nat. Inst. Agric. Sci., Tokyo, Ser. C21,* 111, 1967.

115. **Omar, M. and Heather, W. A.,** Effect of saprophytic phylloplane fungi on germination and development of *Melampsora larica-populina, Trans. Br. Mycol. Soc.,* 72, 225, 1979.

116. **Payghami, E.,** Study of mycoflora of apricot foliage and their antagonism with *Stigmina carpophila* (Lev.) Ellis, causal agent of shot-hole disease, *Iran. J. Plant. Pathol.,* 20, 13, 1984.

117. **Peresse, M. and Picard, D. LE.,** *Hansfordia pulvinata* a fungal parasite destroying *Cladosporium fulvum,* *Mycopathologia,* 71, 23, 1980.

118. **Petrak, F.,** *Sydowia,* 7, 14, 1953, cited in De Vay, J. E., Mutual relationships in fungi, *Annu. Rev. Microbiol.,* 10, 115, 1956.

119. **Pon, D. S., Townsend, C. E., Wessoman, G. E., Schmidt, C. G., and Kingslover, C. H.,** A *Xanthomonas* parasitic on uredia of cereal rusts, *Phytopathology,* 44, 707, 1954.

120. **Porter, C. L.,** Concerning the characters of certain fungi as exhibited by their growth in the presence of other fungi, *Am. J. Bot.,* 11, 168, 1924.

121. **Pruszynska-Gondek, M.,** *Trichothecium roseum* Link ex Fr. on *Uromyces fabae* (Pers.) de Bary, *Acta Mycol.,* 12, 127, 1976.

122. **Purkayastha, R. P. and Battacharyya, B.,** Antagonism of microorganisms from jute phyllosphere towards *Colletotrichum corchori, Trans. Br. Mycol. Soc.,* 78, 509, 1982.

123. **Rahe, J. E., Kue, J., Chuang, C. M., and Williams, E. B.,** Induced resistance in *Phaseolus vulgaris* to bean anthracnose, *Phytopathology,* 59, 1641, 1969.

124. **Rai, B. and Singh, D. B.,** Antagonistic activity of some leaf surface microfungi against *Alternaria brassicae* and *Drechslera graminea, Trans. Br. Mycol. Soc.,* 75, 363, 1980.

125. **Rayss, T.,** Contribution a l'etrude des Deuteromycetis de Palestine, *Palest. J. Bot. Jerusalem Ser.,* 3, 22, 1943.

126. **Reinecke, P. and Fokkema, N. J.,** An evaluation of methods of screening fungi from the haulm base of cereals for antagonism to *Pseudocercosporella herpotrichoides* in wheat, *Trans. Br. Mycol. Soc.,* 77, 343, 1981.

127. **Riggli, J. H. and Klos, E. J.,** Relationship of *Erwinia herbicola* to *Erwinia amylovora, Can. J. Bot.,* 50, 1077, 1972.

128. **Savulescu, T. R. and Sandville, C.,** Quatrie'me Contribution a'la connaissance des Micromycetes de Roumanie, *Anal. Acad. Roman. Mem. Sect. Stintif. Ser. 3,* 115, 397, 1940.

129. **Schreff, R. H.,** Control of bacterial blight of soybean by *Bdellovibrio bacteriovorus, Phytopathology,* 63, 400, 1973.

130. **Sharma, I. K. and Heather, W. A.,** Hyperparasitism of *Melampsora larici-populina* by *Cladosporium herbarum* and *C. tenuissimum, Indian Phytopathol.,* 34, 395, 1981.

131. **Sharma, I. K. and Heather, W. A.,** Antagonism by three species of *Cladosporium* to three races of *Melampsora larici-populina* Kleb., *Aust. J. For. Res.,* 11, 283, 1981.

132. **Sharma, I. K. and Heather, W. A.,** Temperature sensitivity of the antagonism of *Cladosporium* species to races of *Melampsora larici-populina* Kleb. on cultivars of *Populus X euramericana* (Dode) Guinier, *Phytopathol. Z.,* 105, 61, 1982.

133. **Sharma, I. K. and Heather, W. A.,** Post-penetration antagonism by *Cladosporium tenuissimum* to uredinial induction by *Melampsora larici-populina, Trans. Br. Mycol. Soc.,* 80, 373, 1983.

134. **Sharma, J. K. and Heather, W. A.,** Parasitism of uredospores of *Melampsora larici-populina* Kleb. by *Cladosporium* sp., *Eur. J. For. Pathol.,* 8, 48, 1978.

135. **Sharma, J. K. and Heather, W. A.,** Effect of *Cladosporium aecidiicola* Thum. on the viability of urediniospores of *Melampsora medusae* Thum. in storage, *Eur. J. For. Pathol.,* 10, 360, 1980.

136. **Sharma, J. K., Mohanan, C., and Florence, E. J. M.,** Disease Survey in Nurseries and Plantations of Forest Tree Species Grown in Kerala, Kerala Forest Research Institute Research Report No. 36, India, 268, 1985.

137. **Sharma, S. K. and Gupta, J. S.,** Biological control of leaf blight disease of brown sarson caused by *Alternaria brassicae* and *Alternaria brassicicola, Indian Phytopathol.,* 31, 448, 1978.

138. **Sharma, S. K. and Gupta, J. S.,** Role of surface microorganisms of brown sarson (*Brassica* sp.) in relation to *Alternaria brassicae* and *A. brassicicola, Agra Univ. J. Res. Sci.,* 78, 109, 1979.

139. **Sharma, S. K. and Gupta, J. S.,** *Streptomyces rochei* in relation to *Alternaria brassicae* and *A. brassicicola* on the surface of brown 'sarson', *J. Indian Bot. Soc.,* 59, 161, 1980.

140. **Sinha, S.,** Microbiological complex of the phyllosphere and disease control. Presidential Address, *Indian Phytopathol.,* 18, 1, 1965.

141. **Sinha, S. K. and Chaudhuri, K. C. B.,** Control of sorghum anthracnose with *Streptomyces ganmycicus, Indian J. Microbiol.,* 17, 200, 1977.

142. **Sinha, A. K. and Trivedi, N.,** Immunization of rice plants against *Helminthosporium* infection, *Nature,* 223, 963, 1969.

143. **Sleesman, J. P. and Leben, C.,** Microbial antagonists of *Bipolaris maydis, Phytopathology,* 66, 1214, 1976.

144. **Smith, H. S.,** On some phases of insect control by the biological method, *J. Econ. Entomol.,* 12, 288, 1919.

145. **Smith, R. E.,** *Asparagus* and asparagus rust in Calfornia, *Calif. Agric. Exp. Stn. Bull.,* 165, 1, 1905.

146. **Snyder, W. C.,** Antagonism as a plant disease control principle, in *Biological and Chemical Control of Plant and Animal Pests,* American Association for the Advancement of Science, Washington, D.C., 1960, 122.

147. **Spencer, D. M.,** Parasitism of carnation rust (*Uromyces dianthii*) by *Verticillium lecanii, Trans. Br. Mycol. Soc.,* 74, 191, 1980.

148. **Spencer, D. M.,** Parasitism of carnation rust (*Uromyces dianthii*) by *Verticillium lecanii,* in *Annual Report of the Glasshouse Crops Research Institute,* Littlehampton, U. K., 1981.

149. **Spencer, D. M.,** Carnation rust caused by *Uromyces dianthii,* in *Annual Report of the Glasshouse Crops Research Institute,* Littlehampton, U. K., 1983.

150. **Spurr, H. N., Jr.,** Experiments on foliar disease control using bacterial antagonists, in *Microbial Ecology of the Phylloplane,* Blakeman, J. P. Ed., Academic Press, London, 1981, 369.

151. **Stanford, G. B.,** Some factors affecting the pathogenicity of *Actinomyces scabies, Phytopathology,* 16, 525, 1926.

152. **Standford, G. B. and Broadfoot, W. C.,** A note on the biological control of root rots of cereals. Studies of the effects of other soil inhabiting micro-organisms on the virulence of *Ophiobolus graminis* Sacc., *Sci. Agric.,* 11, 513, 1931.

153. **Stewart, D. M. and Hill, J. H.,** An extract produced by *Helminthosporium sorokinianum* toxic to *Puccinia graminis* var. *tritici, Plant Dis. Rep.,* 49, 280, 1965.

154. **Stewart, D. M. and Hill, J. H.,** Toxicity of *Helminthosporium sorokinianum* and other fungi to *Puccinia graminis* var. *tritici, Plant Dis. Rep.,* 49, 371, 1965.

155. **Stolbova, K. A.,** Anti-rust antibiotic activity of strains of *Cladobotryum varium* Nees ex Duby, *Mikol. Fitopatol.,* 117, 343, 1983.

156. **Swell, G. W. F.,** The effect of altered physical condition of soil on biological control, in *Ecology of Soil-borne Plant Pathogens,* Baker, K. F. and Snyder, W. C., Eds., University of California Press, Berkeley, Calif., 1965, 497.

157. **Swinburne, T. R.,** Iron and iron chelating agents as factors in germination, infection and aggression of fungal pathogens, in *Microbial Ecology of the Phylloplane,* Blakeman, J. P., Ed., Academic Press, London, 1981, 227.

158. **Sy, A. A., Norng, K., Albertini, L., and Barrault, G.,** Research on biological control of *Pyricularia oryzae* Cav. III. Effect of temperature on the capacity of antagonistic micro-organisms to inhibit mycelial growth of the parasite *in vitro, Cryptogamie Mycol.,* 4, 245, 1983.

159. **Sy, A. A., Norng, K., Albertini, L., and Petitprez, M.,** Research on biological control of *Pyricularia oryzae* Cav. IV. Effect of pH on the ability of antagonistic microorganisms to inhibit mycelial growth of the parasite *in vitro, Cryptogamie Mycol.,* 5, 59, 1984.

160. **Thomson, S. V., Schroth, M. N., Moller, W. J., and Reil, W. O.,** Efficacy of bactericides and saprophytic bacteria in reducing colonization and infection of pear flowers by *Erwinia amylovora, Phytopathology,* 66, 1457, 1976.

161. **Toronsmo, A. and Dennis, C.,** The use of *Trichoderma* species to control strawberry fruit rots, *Neth. J. Plant Pathol.,* 83 (Suppl. 1), 449, 1977.

162. **Toronsmo, A. and Ystaas, J.,** Biological control of *Botrytis cinerea, Phytopathol. Z.,* 89, 216, 1980.

163. **Trutmann, P., Keane, P. J., and Merriman, P. R.,** Biological control of *Sclerotinia sclerotiorum* on aerial parts of plants by the hyperparasite *Coniothyrium minitans, Trans. Br. Mycol. Soc.,* 78, 521, 1982.

164. **Tsuneda, A. and Hiratsuka, Y.,** Mode of prasitism of a mycoparasite, *Cladosporium gallicola,* on western gall rust, *Endocronartium harknesii, Can. J. Plant Pathol.,* 1, 31, 1979.

165. **Tsuneda, A. and Hiratsuka, Y.,** Parasitization of pine stem rust fungi by *Monocillium nordinii, Phytopathology,* 70, 1101, 1980.

166. **Tsuneda, A., Hiratsuka, Y., and Maruyama, P. J.,** Hyperparasitism of *Scytalidium uredinicola* on western gall rust, *Endocronartium harknessii, Can. J. Bot.,* 58, 1154, 1980.

167. **Tsuneda, A. and Skoropad, W. P.,** Behavior of *Alternaria brassicae* and its mycoparasite *Nectria inventa* on intact and excised leaves of rapeseed, *Can. J. Bot.,* 56, 1333, 1978.

168. **Tsuneda, A. and Skoropad, W. P.,** Nutrient leakage from dried and rewetted conidia of *Alternaria brassicae* and its effect on the mycoparasite *Nectria inventa, Can. J. Bot.,* 56, 1341, 1978.

169. **Van Alfen, N. K.,** Biology and potential for disease control of hypovirulence of *Endothia parasitica, Annu. Rev. Phytopathol.,* 20, 349, 1982.

170. **Van den Heuvel, J.,** Effects of *Aureobasidium pullulans* on number of lesions on dwarf bean leaves caused by *Alternaria zinniae, Neth. J. Plant Pathol.,* 75, 300, 1969.

171. **Van den Heuvel, J.,** Antagonistic Effects of Epiphytic Microorganisms on Infection of Dwarf Bean Leaves by *Alternaria zinniae,* Ph.D. thesis, Willie Commelin Scholten, Baarn, The Netherlands, 1970.

172. **Warren, R. C.,** Interference by common leaf saprophytic fungi with the development of *Phoma betae* lesions on sugarbeet leaves, *Ann. Appl. Biol.,* 72, 137, 1972.

173. **Wibe, O. and Morton, D. I.,** Inhibition of *Septoria passerinii* development in excised barley leaves by *Helminthosporium sativum* and by cell free filtrates, *Phytopathology,* 52, 373, 1962.

174. **Wicker, E. F.,** Natural control of white blister rust by *Tuberculina maxima, Phytopathology,* 71, 997, 1981.

175. **Wollenweber, H. W.,** *Science,* 79, 572, 1934, cited in De Vay, J. E., Mutual relationships in fungi, *Annu. Rev. Microbiol.,* 10, 115, 1956.

176. **Wood, R. K. S.,** Disease resistance in plants, *Proc. Royal Soc. Ser. B, 181,* 213, 1972.

Chapter 2

BIOLOGICAL CONTROL, GENETIC ENGINEERING, AND CROP DISEASE MANAGEMENT

J. E. Rahe

TABLE OF CONTENTS

I. INTRODUCTION

Fifteen years ago, had you asked plant pathologists or other life scientists concerned with crop production about the potential for biological control of crop diseases, the consensus response would have been pessimistic — it did not seem likely that biological agents would ever become widely used for disease control.

The prognosis today appears quite different. A few effective biological agents are currently available commercially. Exploratory and developmental research on biological controls for crop diseases is being undertaken by both governmental agencies and private industrial research laboratories on an unprecedented scale. Governments in the U. S. and Canada are working to establish guidelines for the registration of biologicals. Many scientists believe that crop disease management may soon become a major focus of biological control.

What are the causes of this major shift in outlook in less than two decades? What are some of the new biological controls having potential for use in crop disease management? What are some of the most likely developments for the future?

These questions are addressed here. The first part of this paper reviews three specific biological control subjects which collectively exemplify the emerging application of recombinant DNA technology to higher plants, and give a preview of the potential benefits, risks, and challenges that genetic engineering poses to plant pathologists and other agricultural scientists. The premise that these challenges can best be met when specific controls are viewed within a biologically sound management context is the subject of the remainder of the paper.

II. CATALYSTS FOR A CHANGING OUTLOOK

The reasons for a major change in outlook are several and varied. One significant factor has been a general widening of the concept of biological control. The concept of biological control today embodies not only the introduction of antagonists into cropping systems, but also the "creation" of modified organisms designed to perform particular tasks, and manipulation of the environment to favor resident beneficial organisms via crop rotation, residue management, and a wide range of other cultural practices.[4,12] It is nearly axiomatic today that an introduced antagonist has negligible potential for biological control in an unfavorable environment. Surprisingly, the effect of the environment of the crop ecystem on introduced antagonists, and the effect of crop production methods on resident antagonists do not appear to have been general concerns of much of the earlier research on biological control.

I believe that a second significant factor in these changing attitudes has been the autocatalytic effect of knowledge. It is generally claimed that the first text to be devoted wholly to biological control of plant pathogens was that of K. F. Baker and R. J. Cook,[4] published only 13 years ago. Since then, several texts devoted to this subject have appeared, including a sequel by Cook and Baker.[12] The first text on biological control of plant pathogens represents both the existence and the effective presentation of the critical mass of knowledge required to generate new knowledge on the subject at a significant rate.

A third factor contributing to the increasing interest in biological control is the development of technologies for gene transfer between unrelated organisms, and even the construction of genes. These technologies are changing the world in which we live. Man can now produce "designer organisms", combining the desired characteristics of different natural organisms. The control of crop diseases is but one of the many applications of this technology.

III. CROWN GALL DISEASE

To fan these flames of changing attitudes regarding biological control of plant diseases, the discovery, development, and commercialization of a bacterial antagonist for the biological control of crown gall disease could not have been better timed, even by a Hollywood scriptwriter. Discovered and developed during the early 1970s by Alan Kerr and his associates in Australia,[24,30] *Agrobacterium radiobacter* strain K84 represents the most spectacularly successful biological control yet developed for a plant disease. Of equal or perhaps greater importance to the role of this biological control in contributing to changing attitudes is the fact that both the crown gall pathogen and the biocontrol agent were, and are, important model systems utilized in basic biological research,[14,29] and the physiological and molecular mechanisms of both pathogenesis[5,9,21] and antagonism[23] were elucidated almost simultaneously with the development of the biological control.

Crown gall is a serious disease affecting more than 600 spp. of higher plants.[13] The tumors characteristic of the disease are caused by plasmid insertion into the host plant genome[9,43] The tumor-inducing plasmids are carried by various strains of the soil-inhabiting bacterium *Agrobacterium tumefaciens.*[42] The antagonist, *Agrobacterium radiobacter* strain K84, when applied to plants as a seed or root dip treatment, protects the roots from the tumor-inducing strains by production of a highly selective antibiotic, agrocin 84. Strain K84 is available commercially, commonly as a peat-based formulation, and is applied, usually as a dip or soak treatment, to bare-rooted transplants, cuttings, and graft unions. Strain K84 is used effectively, and on a commercial scale in many countries. It is available under the trade name Dygall from New Zealand, and is registered in the U. S. and marketed as Galltrol. It is not registered for use in Canada, and its use there is illegal.

Almost without question, *Agrobacterium* will have a continuing and enlarged interaction with crop production in the future. This bacterium, a close relative of the nitrogen fixing bacterial genus *Rhizobium*, is one of the most promising and versatile vectors for recombinant DNA transfer into plants.[25] The discovery and development of biological control of crown gall by *Agrobacterium radiobacter* strain K84 has been a most timely and effective catalyst for changing attitudes toward biological control.

IV. ICE NUCLEATING BACTERIA

The discovery of the ice nucleating activity of certain epiphytic leaf surface bacteria, and their contribution to the frost sensitivity of many plant species[27] has also been an effective and timely catalyst. Although physical scientists have known for decades that pure water must be cooled to several degrees below 0°C in order for ice crystal nuclei to be initiated, the phenomenon was believed to have no practical significance to crop production, because it was assumed that ubiquitous dust particles would serve as effective ice nucleators in nature. This assumption was proven false with the discovery that certain epiphytic leaf surface bacteria were effective ice nucleators, and that suppression of these ice nucleating bacteria could provide 4 to 6°C of frost protection for plants under field conditions.[26]

Supression of ice nucleation-active bacteria can be accomplished in several ways. One way is via direct chemical suppression with bacteriocides. A second is via antagonism between ice nucleation-active and -inactive epiphytic bacteria presumably competing for the same leaf surface niches. Lindow and his coworkers[27] have demonstrated the effectiveness of both approaches.

In theory, the best competitor for the leaf surface niche of an ice nucleation-active bacterium should be the same bacterium lacking the capacity for ice nucleation. The genes coding for ice nucleation activity have been identified using techniques of modern molecular biology,[27] and engineered bacterial strains which combine high competitiveness with a lack of ice nucleation activity have been developed.

The past 4 years have witnessed an ongoing political contest in the U. S. concerning the release of genetically engineered ice nucleation antagonists into the environment.[39] A request for the pilot scale commercial field release of one such strain in California, for protection of early season potatoes, was successfully blocked in the courts in February of 1984. A second authorized release scheduled for November of 1985 met local opposition and was blocked by action of a local governmental zoning authority.[41] Genetically engineered biological control agents are presently in limbo. The level of current expenditures by governmental agencies and private research corporations on the development of conventional and genetically engineered biological control agents reflects the confidence of such agencies that their release will be soon forthcoming.

One can only guess at the number of genetically engineered organisms already developed and awaiting economic and political conditions favorable for their release. Among them are ice-minus derivatives of ice nucleation-active bacteria,[39,41] and engineered genotypes of tobacco and tomato with resistance to tobacco mosaic virus.[7]

V. GLYPHOSATE TOLERANT CROPS

Another probable group of engineered plant genotypes awaiting an appropriate climate for release are crop species containing an introduced mutant gene of bacterial origin which confers enhanced tolerance to the herbicide glyphosate. Although these do not exemplify biological control of plant disease, per se, I will describe them here in some detail because they illustrate effectively the potential, the techniques, and some possible risks associated with the development and use of genetically engineered organisms.

A. Nature and Current Uses of Glyphosate

Glyphosate, marketed by Monsanto Corporation under the trade name Roundup, is a remarkable herbicide. It is the phosphonomethyl derivative of the amino acid glycine.[32] It has exceptionally low mammalian toxicity,[6] yet is deadly to most plant species. It is water soluble, easily handled, and is generally claimed to have nonresidual activity in soil,[3,32,35] although some reports[8,15] contradict this latter point. It is absorbed rapidly by foliage and its action results in the eventual destruction of the root system of treated plants. It is one of the few herbicides that can effectively eradicate perennial rhizomatous species such as couch (quack) grass (*Agropyron repens* (L.) Beauv.) and Canada thistle (*Cirsium arvense* (L.) Scop.) with a single application.

At present, glyphosate is used for control of border vegetation around trees, in fencerows and similar situations, for killing sod, as a carefully directed spray for in-row vegetation control in orchards, and for general crop weed control in stale seedbed, minimum tillage, and no-till cropping systems.

Glyphosate has even been used as a component of what can be viewed as a chemical/biological control for broomrape (*Orobranche* spp.), a higher plant parasite of the roots of many broadleaf plant species.[19] In this use, glyphosate "tolerant" species such as broadbeans (*Vicia faba* L.) and carrots (*Daucus carota* L.), sprayed with low doses of glyphosate, can give nearly complete control of shoot emergence of *Orobranche*. Glyphosate treatments resulted in substantial yield increases of broadbeans in *Orobranche*-infested plots,[19] and presumably allowed the broadbeans to act as a trap crop and reduce levels of *Orobranche* seed infestation in the soil.

B. Engineering Specific Tolerance to Glyphosate

Crop varieties with specific tolerance to a water soluble, nonresidual, post emergence herbicide with broad spectrum activity against plants and low mammalian, insect, fish, and aquatic invertebrate toxicity[3,6,16] seem almost too advantageous to be possible. However, engineered plant genotypes with specific tolerance to glyphosate have been, and are being, created in at least some plant species.[10,40]

The primary biochemical target of glyphosate is an enzyme of the shikimic acid pathway.[1,2,38] A mutation causing the production of an altered form of this enzyme and resistance to glyphosate was chemically induced in the bacterium *Salmonella typhimurium*.[11] The mutant gene that produces the glyphosate-insensitive enzyme has been patented under the trade name GlyphoTol.[17] The protein product of the mutant gene, expressed in *Escherichia coli,* was shown to differ from the normal enzyme by a single amino acid substitution which confers a reduced affinity of the enzyme for glyphosate.[37]

C. Crown Gall and Plant Genetic Engineering

The mutant allele was carried into tobacco plants on a modified tumor-inducing (Ti) plasmid from the crown gall pathogen, *Agrobacterium tumefaciens*.[10] In essence, the important modifications of the Ti plasmid were (1) deletion of tumor-inducing activity, (2) addition of resistance to a suitable antibiotic (this resistance permits selection of the plant cells which incorporate the plasmid), and (3) addition of the foreign gene to be transferred (in this case, the mutant gene conferring glyphosate resistance).[10,25] Ti plasmids with modifications (1) and (2) become T-DNA "vectors" for introducing a wide range of "foreign" genes into plant cells. The major advantages of T-DNA vectors are that they permit the transfer and expression of genes from or via bacteria (prokaryotes) to and in plants (eukaryotes). The host range of *Agrobacterium tumefaciens* (the source of Ti plasmids) provides access to a vast number of (potentially all) dicotyledonous plants,[29] as well as to some monocots.[18,25]

T-DNA vectors have already yielded higher plants with specific tolerance to glyphosate,[10] and to tobacco mosaic virus,[7] and they will undoubtably lead to many other genetically engineered crop plant genotypes in the very near future. Thus, a plant disease, and the timely discovery of its unique molecular and physiological nature[9,21,22,42,43] have provided the cornerstone for the application of recombinant DNA technology to higher plants.

It appears likely that authorized controlled testing of one or more genetically engineered organisms will occur in North America in 1986.[40] It is probable that genetically engineered plants and/or microorganisms may be commercially available within 2 or 3 years. If they yield a competitive economic advantage to the user, it is almost certain that they will be used on a vast scale. Engineered crop species with specific resistance to glyphosate will likely be the first of many "designer packages" of chemically dependent crops to be marketed commercially. Such packages have obvious potential for increased ease of crop production. They will permit farming operations to become much larger, and will further increase the tendency of North American agriculture toward area monocropping. They have great potential to contribute to the productivity of the individual "farmer"; whether they will contribute to increased productivity per unit of land area remains to be seen.

VI. GLYPHOSATE RISK ANALYSIS

Any agrichemical, whether used conventionally or as a component of a "designer crop/chemical package", must be used with care and good judgement. Glyphosate-tolerant crop packages will likely lead to a massive increase in glyphosate usage. The published literature already suggests that there are some potential risks associated with the use of glyphosate, and these must be given due consideration.

A. Selection for Tolerance

One obvious risk is that of natural selection for glyphosate-tolerant weeds. This can occur through natural selection for tolerant species,[31] or for tolerant genotypes within a species. Singer and McDaniel[34] reported recently that glyphosate-tolerant cells of tobacco occurred under nonmutagenizing conditions at a frequency of 2.3×10^{-8}, and that this tolerance could be maintained in the absence of selection pressure and was expressed in whole plants regenerated from tolerant cells. If the reported frequency of naturally occurring tolerance in tobacco is typical of that occurring in other plant species, then selection of glyphosate-tolerant weed genotypes in response to the use of glyphosate on a massive scale is inevitable.

B. Direct Effects on Soil Microbial Activity

Glyphosate is not without effect on nonplant species. The shikimic acid pathway occurs in many microorganisms, and these are glyphosate sensitive. The reportedly minimal effects of glyphosate on soil microflora are presumably due to its rapid inactivation in soil.[32,35,36] There are, however, published reports showing that glyphosate does not always have a minimal effect on soil microflora. Bliev[8] observed that organic matter mineralization was inhibited in the field for 445 days after glyphosate applications. Eberbach and Douglas[15] reported that root weight of subterranean clover (*Trifolium subterraneum* L.), its nodulation by *Rhizobium trifolii*, and nitrogen fixation were all strongly suppressed in a sandy loam soil pretreated with a recommended field dose of glyphosate and weathered for 120 days. The data of Reuppel et al.[32] show that while glyphosate was metabolized "rapidly" in two silt loam soils, more than 60% of it was recoverable from a Norfolk sandy loam soil 112 days after application. Total microbial populations in the sandy loam were increased 4- to 6-fold over control by glyphosate treatment, whereas glyphosate had little effect on microbial populations in the finer textured soils. This is intriguing given the fact that it was also shown that microbial degradation of glyphosate was most rapid in the fine textured soils. Sprankle et al.[35,36] also found that glyphosate remained biologically active in sand, but was biologically inactivated rapidly by absorption in clay and loam soils. These reports suggest that the possible effects of glyphosate on general microbial activity and organic matter turnover in soils, and on nitrogen fixation by *Rhizobium* spp. should be further investigated, and carefully considered in any decision to use this chemical.

C. Indirect Effects on Soil Microbial Activity

Glyphosate can also affect soil microflora indirectly. My personal use of Roundup for vegetation control in an orchard situation yielded the repeated observation that accidental contact of the stems of young apple trees with a small amount of the diluted herbicide spray would typically cause negligible symptoms in the year of contact, but often led to reduced growth and vigor in the following year. How could a small amount (typically a few $\mu\ell$-sized droplets) of the diluted herbicide give this kind of delayed effect? This question led to the hypothesis that perhaps a sublethal dose of the chemical promoted the establishment of a deleterious but nonlethal root microflora. This hypothesis, while never tested on apple trees, was proven correct by Johal and Rahe on bean (*Phaseolus vulgaris* L.) seedlings.[20] Glyphosate predisposed the roots of treated bean seedlings to rapid colonization by soil fungi, and this colonization was shown to contribute substantially to the herbicidal activity of glyphosate.

The possible occurrence of this mechanism of action in other plant species is being investigated. The roots of a majority of glyphosate-treated annual and perennial weed species were found to be colonized rapidly by soil fungi, predominantly *Fusarium* spp., under field conditions. Neither the pathogenicity of the root colonizers, nor their possible contribution to the herbicidal effect of glyphosate, has yet been clarified in this ongoing study, but it was shown that glyphosate treatment of established weeds caused significant increases in the number of propagules of *Fusarium* spp. in soil at 3 and 9 weeks after spraying.[44]

D. Decision Making

These various evidences of potential risks associated with the use of glyphosate are not, of themselves, valid bases for objecting to the use of glyphosate or to the release of gly-phosate-tolerant crop genotypes. They are clear indicators of areas where additional research is needed, and of factors that should be considered in relation to the use of this chemical.

These kinds of factors are not unique to glyphosate. They merely exemplify the broad view that must be taken in considering the use of any agrichemical or biological agent, whether genetically engineered or not. Genetic engineering will contribute some spectacular advances to agricultural productivity and to agricultural pest management. Inevitably, it will also contribute to some spectacular problems. Genetic engineering increases tremendously man's ability to simplify resource producing systems. Simple systems are potentially productive, but also potentially unstable. The success of man's exploitation of genetic engineering in the area of agricultural production will depend, not on the genetic engineers, but on the overall wisdom and judgement of the agronomists, biologists, ecologists, industrialists, and politicians who make the decisions on how the powerful tools of this new technology are to be used. In this sense, increasing knowledge of the phenomenon and mechanisms of biological balance will contribute to obtaining the greatest possible benefit from the promise of genetic engineering.

VII. MANAGEMENT OF BIOLOGICAL CONTROLS

Whatever the fate of genetically engineered organisms, experiences with *Agrobacterium radiobacter* strain K84, and antagonists of ice-nucleating epiphytic leaf surface bacteria have clearly shown the immense potential of competitive resident organisms to suppress pathogenic strains of the same or similar organisms in both root and foliar environments. In terms of augmentative or inundative[5] biological control, this will likely prove to be a most significant effect of these two developments of the 1970s. I believe that the short-term future will see the commercial development of a large number of conventional and engineered strains of nonpathogenic but highly competitive organisms with the ability to suppress their pathogenic counterparts in both above- and below-ground environments.

Even so, I believe that the greatest overall potential for an increased role of biological control in crop disease management resides not in commercial biological control agents, but in increasing and exploiting our understanding of the role of various cultural practices on general biological activities in the crop ecosystem.

A. Biological Balance

Why do we have pest problems? Conceptually at least, the answer lies in understanding the nature of biological balance.[4] Nature and natural systems are relatively stable, both despite and because of the fact that individuals and individual species constantly strive for growth, which of itself is destabilizing. Every living species requires and depends upon resources for its existence, and its striving for growth is an effort to bring increasing amounts of resources under its control. As it succeeds it grows, and as it grows it becomes an increasing resource for other living things. As other living things exploit this growing resource, they check its further growth. The system tends toward balance. The greater the number of internal checks on growth of any single component of a system, the greater is its inherent stability.

B. Pest Management or Pest Control?

Man is uniquely equipped among species to bring resources under his control. In most cases he does this by simplifying systems to give maximum resources to the one or more species that he wishes to use as a resource, so that the resource species will increase. In this endeavor, however necessary it may be, man must continually fight the tendency of natural systems to achieve balance. This tendency towards balance takes the form of organisms that continually attempt to exploit the resource that man is attempting to promote. In man's perception, these exploiters are pests, and all too often, merely problems to be eliminated from the natural system. A fundamental weakness of this pest control philosophy, as opposed to a pest management philosophy, is the belief that if a pest can be eliminated the problem that it causes will disappear.

While the belief is true, the implication of no more problems is definitely untrue. Nature tends to utilize available resources. Pest control is aimed at fighting this tendency of natural systems. At best, it can only serve in the long run to replace one unstable system with another. The pest management philosophy recognizes from the outset that if one pest is removed it will in time be replaced by another. Consequently, the pest management philosophy strives to promote a stable complex of organisms, some of which will unavoidably be pests, that can be managed economically to minimize within acceptable limits the damage caused by pests, rather than the pests themselves. The pest management philosophy is to optimize the effect of chemical, genetic, and biological manipulations of the system to provide control of pest damage with maximum long-term stability.

Western agriculture, despite its technological and economic complexities, can be characterized by the word simplification. Land clearing and consolidation; vast areas of crop monoculture, in many instances genetic monoculture; maximal use of readily available, readily lost nutruients; sledgehammer pest control tactics, even to the extreme of burning off the last vestige of remaining organic matter in order to eliminate the pests residing therein; elimination of fence-row and borderland habitats; elimination of natural successional but man-designated weed species of trees in reforestation; and so on — these are some of the obvious characteristics of western agriculture. The economically driven passion of western agriculture for simplification of the crop ecosystem is overwhelming. If it were economically feasible, we would sterilize all of the land on which we grow crops; indeed, much of the California strawberry industry has found annual fumigation of the soil to be a highly profitable practice. We do, and will continue to do, what is economically profitable. Sledgehammer chemical control may never have been biologically sound, but in many cases it was, and is, economically effective, and in some cases, the only viable option available to economically competitive crop production.

Agrichemicals will continue to play a major positive role in both the control and management of crop pests, including diseases. But the increasing costs of disease control chemicals, and the loss in effectiveness of some of these to pest resistance, force us to look to alternatives. Genetically engineered specific biological controls appear to be the alternative of the short-term future.

Specific controls increase man's ability to simplify resource producing systems. Simple resource producing systems are potentially productive, but inherently vulnerable to pests. They can be maintained in a state of quasi stability given a sufficient input of energy to protect the resource being produced. This is feasible for certain high value crops, as exemplified by greenhouse and nursery operations utilizing annually replaced artificial growth media, and certain small fruit and vegetable operations practicing frequent soil fumigation.

But man must also husband the land which produces the vast quantities of lower value staple crops upon which his very existence depends. This environment is manipulated, with or without knowledge, by each and all of the various cultural practices involved in crop production. General biological control has always been operative — it is synonymous with biological balance. Without conscious recognition of this fact, many of the cultural practices of western agriculture have likely served to reduce or eliminate this fundamental component of crop disease management.[4]

Any overall crop disease management program starts with the soil. A major component of biological balance in the soil ecosystem is the level of total microbial activity. The growth of any individual organism in soil reflects the growth promoting effects of available nutrients or resources, and the growth inhibitory effects of competition, metabolic products, and direct attacks by other organisms growing in the same environment. Understanding and exploitation of this phenomenon must be the foundation of any disease management program.

There is abundant indirect evidence to support the concept that the disease producing activities of soil-borne plant pathogens vary inversely with the level of total microbial activity in soil.[4,28] Too little research has been directed to this hypothesis.

If this conceptual relationship is in fact generally operative in nature, then the cultural bases of good soil management and good disease management are one and the same. Total soil microbial activity is maintained at a high level by maximizing the rate of organic matter input and minimizing soil organic matter losses. The rate of input is maximized by utilizing crop residues, making maximal use of winter cover crops, or green manure intercrops wherever possible, eliminating residue burning and other forms of residue loss, and returning organic wastes to the soil wherever economically feasible. Protecting soil organic matter is accomplished by minimizing losses to chemical oxidation and physical losses of topsoil. This in turn is accomplished by minimizing tillage operations and maintaining vegetative cover to the maximum extent possible.

These sound agronomic practices contribute to general microbial activity, and represent the simplest and most fundamental aspects of manipulation of the environment affecting resident beneficial organisms. The effects of these practices are so obvious and yet so frequently ignored. Clearly, there is a great need for much additional fundamental research to be focused in this area.

VIII. CONCLUDING REMARKS

Biologically sound disease management programs must be built upon a foundation of understanding of the role of general soil microbial activity in the phenomenon of biological balance. The next level of biologically sound management can then exploit the selective effects of particular residues and cultural practices in modifying environments to favor particular antagonists.

The last level of biological control is the selective use of commercial agrichemicals, and biological control agents such as *Agrobacterium radiobacter* strain K84, or engineered antagonists of ice-nucleation-active bacterial leaf epiphytes, or plant growth promoting rhizobacteria,[33] or whatever the future holds in store. But at this last level we must not lose sight of the fact that there is biological control and biological management, just as there is pest control and pest management. Without an appreciation of the fundamental role of biological balance in disease management, the only difference between sledgehammer chemical control and the use of commercial biological control agents will be the fact that in the case of biological control the product in the package is alive.

REFERENCES

1. **Amrhein, N., Schab, J., and Steinrücken, H. C.,** The mode of action of the herbicide glyphosate, *Naturwissenschafter,* 67, 356, 1980.
2. **Anton, D. L., Hedstrom, L., Fish, S. M., and Abeles, R. H.,** Mechanism of enolpyruvyl shikimate-3-phosphate synthase exchange of phosphoenolpyruvate with solvent protons, *Biochemistry,* 22, 5903, 1983.
3. **Baird, D. D., Upshurch, R. P., Homesley, W. B., and Franz, J. E.,** Introduction of a new broad spectrum post-emergence herbicide class with utility for herbaceous perennial weed control, *Proc. North Cent. Weed Cont. Conf.,* 26, 64, 1971.
4. **Baker, K. F. and Cook, R. J.,** *Biological Control of Plant Pathogens,* American Phytopathological Society, St. Paul, Minn., 1974, 433.

5. **Batra, S. W. T.,** Biological control in agroecosystems, *Science,* 215, 134, 1982.
6. **Beste, C. E.,** *Herbicide Handbook of the Weed Science Society of America,* 5th Ed., Weed Science Society of America, Champaign, Ill., 1983, 515.
7. **Bialy, H. and Klausner, A.,** A new route to virus resistance in plants, *Bio. Technol.,* 4, 96, 1986.
8. **Bliev, Y. K.,** Influence of glyphosate on the organic matter mineralization and enzymatic activity of the soddy-podzolic soil, *Pochvovedenie,* 4, 74, 1983.
9. **Chilton, M. D., Drummond, M. H., Merlo, D. J., and Sciaky, D., et al.,** Stable incorporation of plasmid DNA into higher plant cells: the molecular basis of crown gall tumorigenesis, *Cell,* 11, 263, 1977.
10. **Comai, L., Facciotti, D., Hiatt, W. R., and Thompson, G., et al.,** Expression in plants of a mutant aroA gene from *Salmonella typhimurium* confers tolerance to glyphosate, *Nature,* 317, 741, 1985.
11. **Comai, L., Sen, L., and Stalker, D. M.,** An altered aroA gene product confers resistance to the herbicide glyphosate, *Science,* 221, 370, 1983.
12. **Cook, R. J. and Baker, K. F.,** *The Nature and Practice of Biological Control of Plant Pathogens,* American Phytopathological Society, St. Paul, Minn., 1983, 539.
13. **De Cleene, M. and De Ley, J.,** The host range of crown gall, *Bot. Rev.,* 42, 389, 1976.
14. **Drummond, M.,** Crown gall disease, *Nature,* 281, 343, 1979.
15. **Eberbach, P. L. and Douglas, L. A.,** Persistence of glyphosate in a sandy loam, *Soil Biol. Biochem.,* 15, 485, 1983.
16. **Folmar, L. C., Sanders, H. O., and Julin, A. M.,** Toxicity of the herbicide glyphosate and several of its formulations to fish and aquatic invertebrates, *Arch. Environ. Contam. Toxicol.,* 8, 269, 1979.
17. **Gebhart, F.,** Calgene obtains first patent for genetically engineered crop gene, *Genet. Eng. News,* 5, 3, 1985.
18. **Hooykaas-Van Slogteren, G. M. S., Hooykaas, P. J. J., and Schilperoort, R. A.,** Expression of Ti plasmid genes in monocotyledonous plants infected with *Agrobacterium tumefaciens, Nature,* 311, 763, 1984.
19. **Jacobsohn, R. and Kelman, Y.,** Effectiveness of glyphosate in broomrape (*Orobranche* spp.) control in four crops, *Weed Sci.,* 28, 692, 1980.
20. **Johal, G. S. and Rahe, J. E.,** Effect of soilborne plant-pathogenic fungi on the herbicidal action of glyphosate on bean seedlings, *Phytopathology,* 74, 950, 1984.
21. **Kerr, A.,** Transfer of virulence between isolates of *Agrobacterium, Nature,* 223, 1175, 1969.
22. **Kerr, A.,** Acquisition of virulence by non-pathogenic isolates of *Agrobacterium radiobacter, Physiol. Plant Pathol.,* 1, 241, 1971.
23. **Kerr, A. and Htay, K.,** Biological control of crown gall through bacteriocin production, *Physiol. Plant Pathol.,* 4, 37, 1974.
24. **Kerr, A.,** Biological control of crown gall through production of agrocin 84, *Plant Dis.,* 64, 25, 1980.
25. **Klee, H. J., Yanofsky, M. F., and Nester, E. W.,** Vectors for transformation of higher plants, *Bio. Technol.,* 3, 637, 1985.
26. **Lindow, S. E.,** Methods of preventing frost injury caused by epiphytic ice-nucleation-active bacteria, *Plant Dis.,* 67, 327, 1983.
27. **Lindow, S. E.,** The role of bacterial ice nucleation in frost injury to plants, *Annu. Rev. Phytopathol.,* 21, 363, 1983.
28. **Marois, J. J. and Mitchell, D. J.,** Effects of fungal communities on the pathogenic and saprophytic activities of *Fusarium oxysporum* f. sp. *radicis-lycopersici, Phytopathology,* 71, 1251, 1981.
29. **Nester, E. W., Gordon, M. P., Amasino, R. M., and Yanofsky, M. F.,** Crown gall: a molecular and physiological analysis, *Annu. Rev. Plant Physiol.,* 35, 387, 1984.
30. **New, P. B. and Kerr, A.,** Biological control of crown gall: field observations and glasshouse experiments, *J. Appl. Bacteriol.,* 35, 279, 1972.
31. **Protopapadakis, E.,** Changes in the weed flora of citrus orchards in Crete in relation to chemical weeding, *Agronomie,* 5, 833, 1985.
32. **Rueppel, M. L., Brightwell, B. B., Schaefer, J., and Marvel, J. T.,** Metabolism and degradation of glyphosate in soil and water, *J. Agr. Food Chem.,* 25, 517, 1977.
33. **Schroth, M. N. and Hancock, J.,** Selected topics in biological control, *Annu. Rev. Microbiol.,* 35, 453, 1981.
34. **Singer, S. R. and McDaniel, C. N.,** Selection of glyphosate-tolerant tobacco calli and the expression of this tolerance in regenerated plants, *Plant Physiol.,* 78, 411, 1985.
35. **Sprankle, P., Meggitt, W. F., and Penner, D.,** Rapid inactivation of glyphosate in soil, *Weed Sci.,* 23, 224, 1975.
36. **Sprankle, P., Meggitt, W. F., and Penner, D.,** Adsorption, mobility and microbial degradation of glyphosate in the soil, *Weed Sci.,* 23, 229, 1975.
37. **Stalker, D. M., Hiatt, W. R., and Comai, L.,** A single amino acid substitution in the enzyme 5-enolpyruvylshikimate-3-phosphate synthase confers resistance to the herbicide glyphosate, *J. Biol. Chem.,* 269, 4724, 1985.

38. **Steinrücken, H. C. and Amrhein, N.,** The herbicide glyphosate is a potent inhibitor of 5-enolpyruvyl-shikimic acid-3-phosphate synthase, *Biochem. Biophys. Res. Commun.*, 94, 1207, 1980.
39. **Sun, M.,** EPA approves field test of altered microbes, *Science*, 230, 1015, 1985.
40. **Sun, M.,** Engineering crops to resist weed killers, *Science*, 231, 1360, 1986.
41. **Sun, M.,** Local opposition halts biotechnology test, *Science*, 231, 667, 1986.
42. **Watson, B., Currier, T. C., Gordon, M. P., and Chilton, M. D., et al.,** Plasmid required for virulence of *Agrobacterium tumefaciens*, *J. Bacteriol.*, 123, 255, 1975.
43. **Yadav, N. S., Postle, K., Saiki, R. K., and Thomashow, M. F., et al.,** T-DNA of a crown gall teratoma is covalently joined to host plant DNA, *Nature*, 287, 458, 1980.
44. **Levesque, A. and Rahe, J. E.,** unpublished data.

Chapter 3

PLANT BREEDING STRATEGIES FOR BIOLOGICAL CONTROL OF PLANT DISEASES

Gurdev S. Khush and S. S. Virmani

TABLE OF CONTENTS

I. INTRODUCTION

Plant species are subjected to attack by many diseases caused by diverse pathogens — fungi, bacteria, viruses, mycoplasmas, and nematodes. Breeding is the most widely used and most effective method used to control plant disease. Disease resistance per se does not improve yield potential; it merely keeps the crop healthy, allowing it to express its full performance potential.[10] Resistance breeding involves genetic manipulation of the host-parasite interactions. A basic feature of this interaction is the fact that the parasite must circumvent or resist the defense mechanisms of the host while the host must likewise avoid or resist the parasite to a certain extent. Plants have evolved complex mechanisms of resistance to diseases, which at best are only partially understood at present. Though the genetics of resistance is known, the mechanisms by which resistant reactions are brought about remain unknown in many cases.

According to Hooker,[11] recorded observations about various reactions of cultivated crops to diseases go back to the time of Theophrastus (371-286 BC). T. A. Knight, an English plant breeder, noted rust resistance in wheat. Breeding programs in potato for late blight resistance were under way after 1851, and had the support of Charles Darwin.[11] In France, breeding of grapes for downy-mildew resistance was started in 1878. Aside from these instances, there were relatively few cases of deliberate breeding for plant disease control. Breeding for resistance in plants received great impetus after Biffen[1] showed that resistance to disease was an inherited character. In the past 80 years, tremendous progress has been made in breeding crop plants for disease resistance. Today resistance breeding is a major component of any crop improvement program.

This paper highlights some plant breeding strategies used to control plant diseases.

II. DURABILITY OF RESISTANCE

Two types of resistances are encountered by plant breeders: (1) the highly unstable incompatibility genes often expressed as a hypersensitive or low infection type reaction to a biotrophic pathogen, and (2) the more durable resistance genes that affect the basic compatibility processes in host-pathogen relationships. Examples of the unstable incompatibility genes in crop species are wheat (*Ustilago nuda* f. sp. *tritici aestivi*), rice (*Pyricularia oryzae*), oats (*Puccinia coronata*), flax (*Melampsora lini*), tomato (*Clasdosporium fulvum*), potato (*Phytophathara infestans*), and maize (*Puccinia sorghi*). All pathogens involved in these interactions are biotrophic. It is often difficult to classify resistances that affect basic pathogenicity processes with certainty because so little is known about the mechanisms by which they operate.[11] To use durability of a resistance gene as the basis of classification can be erroneous because incompatibility genes differ in durability.[7]

However, it is helpful if the plant breeder knows whether he is dealing with unstable or durable resistance. The best guide in this situation is past experience. Resistance, once shown to be durable, can be introduced and selected in the same way as other traits are incorporated into new cultivars. To protect crops by means of incompatibility genes requires a different strategy. These genes need "protection" so they must be combined with other genes such that combination interferes with the development of new adapted races of the pathogen.[11]

III. BREEDING STRATEGIES

Several breeding strategies have been used for disease control depending upon the types of sources of resistance available and the nature of the pathogen. Some of the strategies used are as follows.

A. Absence of a Strategy

The most common strategy used in breeding a crop for disease resistance actually appears to be no strategy at all.[15] The primary objective is to develop good resistant cultivars and often this results in most cultivars being developed for a given growing area and carrying the same resistance gene because it represents the best source of resistance. The breeding for barley resistance to powdery mildew, wheat to yellow rust, and rice to bacterial blight, are some examples of this strategy. The situation is characterized by a fairly rapid turnover in cultivars used where only a few race specific resistances are exposed to the pathogen population at any given time. The most common reason for a cultivar replacement is because it has become susceptible to a new race of a different pathogen.

B. Sequential Release

This strategy uses one resistance gene in agricultural production of the crop at any given time, and the virulence-gene composition of the pathogen population, especially relative to the resistance gene being used, is monitored annually on a differential series of host genotypes that carry different resistance genes singly or in various combinations. Thus, a breeder can detect changes in the frequencies with which virulence genes occur, especially the increases in the frequencies of those that can parasitize the cultivar which carries the resistance gene currently in use.[15] As soon as a new race that is virulent on the currently used resistance gene appears, new cultivars that carry another effective gene are released. This kind of breeding strategy tries to keep a step ahead of the pathogen. Good examples are the breeding of flax for flax rust resistance in North America, and rice for tungro resistance at the International Rice Research Institute (IRRI) in the Philippines. From the 1930s to date, flax has been protected from flax rust by a succession of resistant cultivars. Consequently, the resistance genes L9, P, M, L, and N 1 were introduced and each broke down to a new race; their period of effectiveness ranged from 5 to 13 years. Presently, resistance genes L6 and L11 are in use.[8,19]

C. Gene Pyramiding

This strategy involves the placement of two, three, or more new and still effective resistance genes into a new cultivar so that the pathogen population has a barrier of several resistances presented to it simultaneously. This should be an effective strategy because a new race, to overcome the multiple resistance genes, must have two or three simultaneous changes towards virulence. However, a new race needs only one virulence gene change to overcome a single gene for resistance.[8] This strategy can operate satisfactorily only when breeding is centrally coordinated, and when the production area is isolated from other areas where the system is not applied. If some cultivars with 2 to 3 pyramided genes and the others with single genes are released simultaneously, the effectiveness of this approach will be reduced materially.[15]

Stem rust resistance in Australia has been controlled through this strategy. During the 1930s, 1940s, and early 1950s, the genes Sr6, Sr11, Sr9b, and SrTt were used singly in succession to protect wheat from the rust pathogen. But since the late 1950s, two or more Sr-genes have been pyramided into the same cultivars, and at present, the breeding and selection for multiple gene resistance is well coordinated.[18]

D. Polygenic Resistance

When the virulence system of the pathogen is under polygenic control, the polygenes have to be accumulated in a single genotype of a cultivar to provide a multiple-resistance barrier against the pathogen. Usually, it is not possible to concentrate all of the required genes into one genotype in a single breeding cycle. Therefore, a breeding method involving recurrent selection is used so that the improvement of resistance is a continuous process.

Examples are resistance to *Phytophathara infestans* in potatoes,[2] leaf rust in barley,[16] *Puccinia sorghi* in maize,[9] and downy-mildew in *Pennisetum americanum*.[20]

E. Gene Rotation

This strategy of plant disease control involves planting of resistant genotypes in rotation. The rotation of cultivars with major genes for resistance is based upon the concept that races of a pathogen indigenous to a cropping area are a result of the presence of major genes for resistance in varieties being grown in the area.[4] The primary advantage of the strategy of rotating cultivars with major genes for resistance is that genes are deployed only when needed, and a means of conserving and reusing genes is established. The strategy assumes that new genes for pathogenicity will arise in the pathogen population, and the strategy is designed to control stress before they become important.[5] Yield losses due to the new races will be minimized and the super races will be prevented from developing. This strategy is highly amenable to breeding programs concentrating on the development of F_1 hybrids as varieties.

To be effective, the strategy requires a vigorous plant breeding and crop variety development program. It is being used successfully to control rice blast in Korea and with certain high value crops in North America.[6]

F. Multilines and Cultivar Mixtures

A multiline variety is made up of several component lines. Each line is as genetically similar as possible to every other line, except for major genes for resistance to a specific pathogen. Browning and Frey[3] discussed in detail the use of multiline varieties as a disease control strategy in oats. The major advantage of multiline varieties is that they confront the pathogen with several major genes for resistance simultaneously.

Each major gene for resistance is present in a separate component line. Multilines provide insurance against total crop failure because some of the components should be resistant to certain pathogen races. Besides, the component lines with the individual genes for resistance have a dilution effect on the inoculum potential.[6]

Crill et al.[6] also suggested the possibility of using F_1 hybrid varieties in a disease control strategy with multiline varieties. If the major genes for resistance are dominant, they need to be present in only one parent — either the pollen or seed parent. If the seed parent is selected as common for all components of the F_1 hybrid multiline, the individual major genes may be incorporated into the common pollen parent through backcrossing. The pollinator lines would not have to be nearly as uniform as when used directly as varieties. After such a series of lines is established, it should be possible to generate new multiline varieties of F_1 hybrids quite rapidly, and at the same time improve yield potential and quality.

The objection to the use of multiline varieties has been the time required to develop component lines, and their nonuniformity. A theoretical disadvantage is that the deployment of several major genes for resistance at one time generates selection pressures for the formation of a super race, which would cause all components of the multiline to be come susceptible. This should not occur if pathogen race surveys are conducted and only component lines resistant to the prevalent races are included.

G. General Vs. Specific Resistance

General resistance is resistance against all races of pathogens; such resistance has also been referred to as durable resistance[12] or horizontal resistance.[17] It is usually, but not always, polygenic in inheritance. Breeders and pathologists have found that a moderate level of general resistance is often adequate to prevent field losses.[14] The most important feature of general resistance is that it does not put the same type and degree of selection pressure on the pathogen as does specific resistance. All races are kept at a low level and little disease develops. The resistance seems to remain effective indefinitely. Whether or not a resistance being employed in a cultivar is "general" can be determined by wide area testing, preferably on a multilocation basis and over many seasons. A form of resistance that has functioned well for 20 years or more is likely to be a general type.[11]

There are practical difficulties in incorporating this type of resistance into the improved germplasm. The resistant donors are land varieties and generally have poor plant types. In the process of plant selection for better agronomic traits in crosses involving these donors, the resistance levels are lost or diluted. It is extremely difficult to identify genotypes with resistance for the disease because the precise screening techniques for identifying such genotypes in segregating breeding populations have not been developed.[13] Another difficulty is the length of time required for developing varieties with general resistance. At least 10 to 12 years would be needed to accumulate enough polygenes from several donor parents to build sufficient levels of resistance. When multiple disease resistance is required, as in the case of rice in the tropics and subtropics, it is next to impossible to develop multiple-disease resistant varieties possessing a good level of polygenically controlled general resistance to each disease.[13] Even if a variety with an acceptable level of resistance for a particular disease is developed, it may not stay in production for a sufficient number of years to justify the time and resources taken to develop it. It may go out of production because it may succumb to another disease or insect, or simply because of the changing needs of a dynamic agriculture.

H. Resistance Genes in Integrated Disease Control

To combat the most serious pathogens, resistance genes should be employed as a component of a completely integrated disease management program involving proper agronomic and phytosanitary measures, and chemical control of the pathogen to reduce its population. The useful life of a major gene would thus be prolonged.

IV. CONCLUSIONS

Although many breeding strategies for developing disease resistant cultivars have been proposed and tried, the most commonly deployed strategy is the sequential release of cultivars with major genes for resistance. Gene pyramiding is rarely employed because it takes time, and one needs a collection of races to identify genotypes with a specific combination of genes. Aside from gene pyramiding for wheat stem rust resistance in Australia,[18] there are a few other examples of using this approach. The sole example of the use of multiline varieties for disease control are the multiline varieties of oats against crown rust.[3] Similarly, gene rotation has been practiced in only a few cases.[6] There are many practical difficulties which hinder the progress in developing varieties with polygenic or general resistance.[13] Under these circumstances, the sequential release of cultivars with diverse genes, combined with integrated disease management, appears to be the best strategy.

REFERENCES

1. **Biffen, R. H.,** Mendel's laws of inheritance and wheat breeding, *J. Agric. Sci.,* 5, 4, 1905.
2. **Black, W.,** The nature and inheritance of field resistance to late blight (*Phytophathata infestans*) in potatoes, *Am. Potato J.,* 47, 279, 1970.
3. **Browning, J. A. and Frey, K. J.,** Multiline cultivars as a means of disease control, *Annu. Rev. Phytopathol.,* 7, 355, 1969.
4. **Crill, J. P.,** An assessment of stabilizing selection in crop variety development, *Annu. Rev. Phytopathol.,* 15, 185, 1977.

5. **Crill, J. P. and Khush, G. S.,** Effective and stable control of rice blast with monogenic resistance, *Ext. Bull. 128, Food and Fertilizer Technology Center,* Taipei, Taiwan 1979, 13.

6. **Crill, P., Nuque, F. L., Estrada, B. A., and Bandong, J. M.,** The role of varietal resistance in disease management, in *Evolution of the Gene Rotation Concept for Rice Blast Control,* International Rice Research Institute, Los Banos, Laguna, Philippines, 1982, 103.

7. **Eenink, A. H.,** Genetics of host-parasite relationships and uniform differential resistance, *Neth. J. Plant Pathol.,* 82, 133, 1976.

8. **Flor, H. H. and Comstock, V. D.,** Flax cultivars with multiple rust-conditioning genes, *Crop Sci.,* 11, 64, 1971.

9. **Hooker, A. L.,** Widely based resistance to rust in corn, *Field Crops. Spec. Rep. 1a, Agric. Home Econ. Exp. Stn.,* 64, 28, 1969.

10. **Hooker, A. L.,** A plant pathologist's view of germplasm evaluation and utilation, *Crop Sci.,* 17, 689, 1977.

11. **Hooker, A. L.,** Breeding to control pests, in *Crop Breeding,* Wood, D. R., Ed., American Society of Agronomy, Madison, Wis., 1983, 199.

12. **Johnson, R.,** Durable resistance: definition of genetic control and attainment in plant breeding, *Phytopathology,* 72, 567, 1981.

13. **Khush, G. S. and Virmani, S. S.,** Breeding rice for disease resistance, in *Progress in Plant Breeding I,* Russel, G. E., Ed., Butterworths, London, 1985, 239.

14. **Nelson, R. R., Ed.,** *Breeding Plants for Disease Resistance, Concepts and Applications,* Pennsylvania State University Press, University Park, Pa., 1973.

15. **Parlevliet, J. E.,** Disease resistance in plants and its consequences for plant breeding, in *Plant Breeding, II,* Frey, K. J., Ed., Iowa State University Press, Ames, Iowa, 1981.

16. **Parlevliet, J. E. and Van-Ommeren, A.,** Partial resistance of barley to leaf rust, *Puccinia hordei.* II. Relationship between field trials, micro-plot tests and latest period. *Euphytica,* 24, 293, 1975.

17. **Van der Plank, J. E.,** *Disease Resistance in Plants,* Academic Press, New York, 1968, 206.

18. **Watson, I. A.,** *The National Wheat Rust Control Program in Australia,* University of Sydney, Sydney, Australia, 1977, 24.

19. **Zimmer, D. E. and Hoes, J. A.,** Race 370, a new and dangerous North American race of flax rust, *Plant Dis. Rep.,* 58, 311, 1974.

20. **Gill, K. S.,** Director of Research, Punjab Agricultural University, Punjab, India, personal communication.

Chapter 4

APPLICATION OF BIOLOGICAL CONTROL AGENTS

James P. Stack, Charles M. Kenerley, and Robert E. Pettit

TABLE OF CONTENTS

I. INTRODUCTION

The application of biological control agents to the environment for the control of plant diseases (e.g., field, microplot, orchard, or glasshouse) encompasses the preparation and delivery of the agents, methods for the evaluation of the efficacy of the applied agents, as well as consideration of the potential impact the agents may have upon nontarget organisms. Recent reviews covering these topics have been published.[20,37,46,65,73] The intent of this chapter is to discuss principles believed by the authors to be important to the application of biological control agents, rather than to review the entire body of literature. Because of our current research interests with the biological control of soil-borne pathogens, most examples presented will be for soil-borne pathogens. The approach and principles discussed here will be appropriate to the development of biological controls of both foliar and soil-borne plant pathogens. The development of screening strategies for selection of biocontrol agents is the same for foliar or soil-borne plant pathogens.[6]

It is our opinion that to develop effective biological controls for plant pathogens, it is counterproductive to consider the components of a biocontrol system as independent units (e.g., selection of the agent, mechanism of control, delivery of the agent). During the development of biological controls, a consideration of all components and the integration of the components must be made. Consequently, the emphasis of this chapter will be to discuss an approach to the development of application procedures based upon how the application of biocontrol agents integrates with a total biological control system.

II. AN APPROACH

There may be much to be gained by thinking of biocontrol agents in the same manner as we think about plant pathogens. In fact, many biocontrol agents are pathogenic to the target propagule, causing death and lysis of cells and loss of viability of multicellular propagules (e.g., sclerotia). With plant pathogens, we try to learn as much about their life history as possible in order that we might manipulate it to the hosts advantage. By the same logic we should try to learn as much as possible about the life history of biocontrol agents in order that we might manipulate them to our advantage, which in this case would be to the hosts' (plant pathogen) disadvantage. In both cases we need a thorough understanding of the epidemiology of the diseases. Information on the production and dispersal of inoculum, the inoculation, infection, and colonization processes, as well as reproduction and survival is lacking for most biological control agents.

When conducting field research with plant pathogens, great effort is afforded to creating conditions conducive to disease development in order to get uniform infection and high levels of disease. There are certain criteria which must be met with respect to attributes of the pathogen and environmental conditions. This is precisely what is needed for field research with biocontrol agents. We need to establish conditions favorable to an epidemic of the biocontrol process. This means determining what environmental factors influence agent performance and the intended biocontrol, and then attempting to regulate the environment, for example, through timely irrigations and agent delivery, to optimize chances for success. This is rarely done.

If biocontrol technology is to be transferred to producers it must be cost effective. The economics of biocontrol will ultimately determine the magnitude of its utilization. Involvement of agricultural engineers and agricultural economists in the development process might be a positive influence on the direction and scope of the biocontrol program. Agricultural engineers could help in adapting existing technology or developing new technology for the application of the biocontrol agent. Agricultural economists could help determine the probable costs and anticipated return on investments for biological controls alone or integrated with other control measures, as well as assist in determining the most appropriate markets.

For the most part, the development of application procedures for biocontrol agents has been driven by trial and error rather than motivated by an understanding of the plant-plant pathogen-biocontrol agent system. The first stage in the development of a biological control program should be to define the objective, i.e., to determine the task to be accomplished by the biocontrol agent. This is dependent upon the strategy or strategies of control,[21] whether it be to protect the host,[22] to directly and adversely affect the pathogen population,[3] to replace the pathogen with saprophytes,[16] or to stimulate host resistance mechanisms.[77,78] How the strategy relates to the plant host-plant pathogen interaction should be considered. For example, if the strategy is host protection, then the length of time protection is required should be taken into account.

Having identified the task, the combination of attributes the agent must have to accomplish that task should be determined. For example, if the objective is to directly and adversely affect target propagule germination and/or viability in the field, then the desirable attributes may include an ability to reach and initiate contact with a target propagule, as well as, an ability to affect germination and viability. The ability to reach and make contact with the target propagule will vary depending upon where the target propagules are found, (free in soil or embedded in organic matter, on the leaf surface, or under bud scales) and whether the agent can be delivered to that site (e.g., leaf surfaces) or needs to reach the infection court through its own efforts (e.g., target propagules buried in soil).

Next, one needs to develop a screening assay which would allow the selection of agents based upon the desired attributes.[45] This phase in the development process is critical to the success of the total program. A thorough and thoughtful presentation by Andrews on selection strategies is recommended reading.[6] Having selected potential agents, one should try to determine the mechanism by which control is being effected. If we do not understand the mechanism of control, then how can we utilize it most efficiently?[8] The development of biocontrols should be a dynamic process. As information is gained, the approach should be reassessed for its potential to fullfill the stated objectives. For example, as the mechanisms of control are elaborated, modifications of the presumed desired attributes may need to be made. Determining the relative importance of agent attributes should in turn enhance the chance of further elucidating the actual mechanism of control.

With a preliminary understanding of the mode of action and a list of presumed desirable attributes, it is next to determine what factors (e.g., environmental, nutritional) influence the performance of the agents and which of these factors can be manipulated to maximize the agents performance. For example, by regulating substrate composition can we elicit a desired agent response, or by timely irrigation can we extend the duration of the agent's activity? It is at this point, based upon all of the above, that selection and testing of the method of application should be done. The method of application should be appropriate to the task to be accomplished, the desirable agent attributes, the mechanism by which control is to be effected, and the factors which influence the performance of the agent.

Most biological controls considered to date have involved the application of a single biological control agent. The simultaneous or sequential introduction of more than one agent to control a pathogen or protect a plant host may offer additional opportunity for successful and durable control. Chemical treatment of seeds often involves more than one pesticide. Seed coating with biocontrol agents could also involve more than one agent. Adding a combination of biocontrol agents may be necessary to override the specificity observed with some mycoparasite-plant pathogen interactions.[39] Also, some organisms may be effective only in the presence of other organisms. There is some indication that *Bacillus* spp. and *Clostridium* spp. act in concert to significantly reduce the survival of *Peniophora sa crata*.[79] No such effect was observed when the bacterial species were applied singly.

As with chemicals, compatibility studies would be required. Theoretically it would be possible to treat a soil with an agent to directly and adversely affect the plant pathogen population (e.g., soil-borne root pathogens, overwintering pathogens of above ground plant parts), coat the plant seed with an agent to protect the host (e.g., root or hypocotyl pathogens, damping-off organisms), and at some future time of plant vulnerability to apply an agent to a specific plant organ (e.g., leaf, fruit) for protection or eradication of a pathogen.

III. NATURE OF SUBSTRATE

As stated earlier the economics and mechanics of application will ultimately determine which system is utilized. Hence, agricultural engineers and agricultural economists should be involved, or at least consulted, during the development of biological controls.

In terms of designing application procedures to elicit specific agent responses, it may be profitable to consider substrate and carrier independently. In some cases the carrier and the substrate are the same, for example wheat straw[2] and beet pulp.[58] The straw is both the physical unit upon which the agent is delivered and the chemical source of energy. In other cases they are separate components; for example, lignite granules and thin liquid stillage.[42] The lignite acts as the carrier, but the stillage is the source of energy. There may be more flexibility in a system where the carrier and substrate are independent, thus providing an opportunity for manipulation.

For many plant pathogens, virulence is a function of pathogen nutrition.[40] It is probable that virulence of biological control agents towards plant pathogens is also a function of nutrition. The importance of a food base for the efficacy of a biocontrol agent has been reported.[21,37,65,74] There are few cases where the mechanisms of control are understood. Some of the proposed mechanisms for biocontrol are antibiotic production,[25,67] synthesis and excretion of cell wall degrading enzymes,[29] mycoparasitism,[5,9,39,53] and competition.[53] The synthesis of antibiotics by bacteria and fungi is under nitrogen regulation and subject to catabolite repression.[4,55,63] The form and concentration of nitrogen and carbon is important to the kinetics of antibiotic synthesis and perhaps to the excretion of the compounds as well. If the effectiveness of a biocontrol agent is tied to the synthesis of antibiotics, then clearly the substrate upon which the agent is grown is a crucial component of the biocontrol system. The synthesis of cell wall degrading enzymes is under similar regulatory processes and their production will be a function of the substrate upon which the agent is grown. Under nutrient-limiting conditions, the synthesis of catabolic enzymes (e.g., cell wall depolymerizing enzymes) may be decreased.[38]

Therefore, it is important not only to provide a substrate that will promote the synthesis of the desired enzymes, but also to provide sufficient substrate so as not to limit the synthesis of the enzymes at the time they are required. Another possibility is to create regulatory mutants of the biocontrol agents which will not be subject to catabolite repression and perhaps not require an inducer. Such regulatory mutants of catabolic enzymes (e.g., β-galactosidase) do occur, at least in bacteria.[61] This would allow for continuous production of antibiotics or enzymes during the growth of the agent, providing the substrate was appropriate.

The composition and concentration of the substrate is likely to be important to mycoparasitism and competition as well. The importance of the concentration and source of carbon and nitrogen to the growth of biocontrol agents in soil has recently been shown.[75] The percentage of the carrier granules from which hyphal growth into soil occurred, the number of hyphae extending into the soil from each carrier granule, and the actual length of the hyphal strands were all affected by the composition, concentration, and carbon to nitrogen ratio of the substrate. It was observed that certain substrates were good for achieving a high degree of agent colonization of carrier granules during the preparative stage, but were poor relative to the performance of the agent in soil. Although 100% of the granules were colonized, a very low percentage supported hyphal growth of the agent into the soil.

The best substrates for any biocontrol agent will depend upon the mechanism of control and the other desirable agent attributes. It is therefore likely that the optimal substrate (composition and concentration) will vary with the system in which control is to be effected.

IV. NATURE OF AGENT PROPAGULE

Another factor to consider is the agent propagule itself. Recent work has shown that mycelial preparations or germinated conidia of *Trichoderma harzianum* and *Gliocladium virens* had significantly greater biocontrol capabilities than nongerminated conidia when added to soil.[50,51] To increase the survival potential of the agent, there may be further advantages to using more durable propagules such as chlamydospores.[50-52] However, whether the chlamydospores resulted in an acceptable level of disease control was not stated in these reports. This needs to be determined since it is clear that different propagules have different capabilities and just the presence of the agent alone is not sufficient to effect control.

V. METHODS OF DELIVERY

A. Drench

There are several ways to ammend soil with a biocontrol agent. The easiest and most direct approach is to drench the soil with a suspension of agent propagules.[2,56] Depending upon the medium (e.g., natural soil or nonsoil mix) in which control is being attempted, there are a few potential limitations to this approach. The first is the inability to get a thorough infiltration of heavy-textured soils; the propagules of the agent can be filtered by the soil and hence diluted with increasing depth.[81] It has been shown in undisturbed soil cores that passive dispersal through the soil profile during infiltration occurs chiefly through channels and cracks, not through the pore system.[71] It would therefore be difficult to achieve an even distribution. If the target propagules to be affected, or the plant part to be protected, are near the soil surface, then the effect of concentrating the agent in this zone may be an advantage. If on the other hand, the target propagule to be affected or the plant surfaces to be protected are distributed throughout the soil profile, the concentration of agent propagules in the surface horizon could be a disadvantage. A second limitation may be in adding the agent to the environment in a manner that gives it no competitive advantage against the indigenous microflora for the available substrate. Agent propagules, independent of a protective carrier, may be subject to fungistasis and not germinate although they remain viable.[54] Also, exposed propagules may be more susceptible to predators and parasites.

B. Colonized Natural Substrates

Numerous attempts have been made to control several soil-borne pathogens by incorporating natural substrates colonized by antagonists of the pathogens into soil. The results of these attempts have varied according to substrate. Lowered disease severity, increased yield, and a decreased pathogen population (*Rhizoctonia solani*) resulted from incorporation of wheat bran colonized by *Trichoderma harzianum* in strawberry nursery and field plots.[30] *Rhizoctonia solani* on beans and carrots,[76] tomatoes and peanuts,[31] potatoes,[33] and iris[18] has been controlled to at least some extent by the incorporation of *T. harzianum*-colonized wheat bran into the plant growth medium. *Pythium aphanidermatum* damping-off of beans, cucumbers, peas, and tomatoes was controlled by a *T. harzianum*-colonized wheat bran/peat mixture.[70] Although wheat bran has been effective in many cases, it was ineffective in controlling Rhizoctonia damping-off of radish seedlings.[60] *Laetisaria arvalis*-colonized corn leaf meal or sugar beet pulp prevented an increase in the *Pythium* spp. populations in nonsterile soil.[58]

In this case the noncolonized corn leaf meal stimulated the *Pythium* populations. It would therefore seem very important to insure a thorough colonization of the substrate by the biocontrol agent, as any noncolonized substrate may aid the plant pathogen.

In a comparison of natural carrier/substrates, it was reported that *T. harzianum*-colonized barley grain was much more effective than colonized wheat straw or bean straw in controlling white rot of onion (*Sclerotium cepivorum*).[2] This clearly illustrates the importance of the substrate and carrier system in the ultimate performance of the biocontrol agent. The application procedure should not be evaluated just in terms of the mechanics of delivery; subsequent agent performance may be a function of the substrate and carrier system utilized.

Incorporation of composted hardwood bark into container media has given control of Rhizoctonia damping-off of raddish seedlings.[64] The efficacy was associated with *Trichoderma* spp. that naturally colonize the compost medium.[48] Deliberate and controlled infestation of hardwood bark with select biocontrol agents may be a profitable approach to pursue.

C. Colonized Synthetic Substrates

Embedding propagules of biocontrol agents in a matrix formed by sodium alginate and clay appears to be an effective method of delivery for *Gliocladium virens*, *Penicillium oxalicum*, *Talaromyces flavus*, and *Trichoderma viride*, but not for *Pseudomonas cepacia*.[35] In this system, a mixture of propagules and the matrix are formed into pellets which can then be applied to the growth medium (e.g., soil). Certain agents survived longer than 12 weeks in the pellets.[35] Incorporation of bran into the alginate pellets, as the bulking agent instead of the inert clay Kaolin, resulted in a greater proliferation in soil of *Trichoderma viride*, *T. hamatum*, *T. harzianum*, and *Gliocladium virens*.[50] Other promising methods are based upon inert granules such as diatomaceous earth[10] or lignite.[42] The lignite, which can be ground to specific size granules, and the diatomaceous earth are easily impregnated with a substrate for growth of the biocontrol agent. After colonization by the agent, the granules can be applied to the growth medium either in a glasshouse or the field. As with the alginate pellets there is good survival potential by the agents on lignite granules. With lignite, the performance of the agent is very much dependent upon the available substrate.[75] Whether the performance of agents in alginate pellets could be enhanced by supplementing with specific nutrients should be tested.

D. Aerial Sprays

Control of foliar pathogens and insects has been achieved by spraying agents directly onto the above ground plant parts.[17,26,36,66,72] Physical factors which affect the efficacy of these agents appear similar to those which affect the efficacy of foliar pesticides; namely, spray rate, spray schedule, droplet size, and carrier liquid.[26,72]

Unlike the soil environment, which offers some buffering capacity, the physical environment of the phylloplane can be harsh and subject to rapid and extreme variation in moisture and temperature. Consequently, survival of applied biocontrol agents has generally been poor.[26,49,74] Because of the rapid decline in the population of *Chaetomium globosum* applied on the leaf surface, it was necessary to reintroduce it every 1 to 2 weeks to achieve control of the apple scab pathogen *Venturia inaequalis*.[26] Control of *Gremmeniella abietina* on pine seedlings was achieved when a strain of *Pseudomonas fluorescens* was applied in a dilute nutrient solution.[74] No control was obtained when the bacteria were added in water. Survival of *P. fluorescens* on seedlings was also greater when the bacteria were applied in nutrient solution than when applied in water. In light of the poor survival in the phylloplane,[49] more attention should be given to the role the delivery system and the available substrate play in agent survival and subsequent performance.

E. Other Methods

Control of plant diseases by applying biological agents to the infection court has been reviewed recently.[46] Success has been achieved for control of a number of pathogens with the application of fungal agents[18,31,32,34,35,41,68,82] or bacterial agents[22,80,82] to bulbs, seeds, potato seed pieces, or harvested fruits. This approach was also successful for control of verticillum wilt of tomato by a root dip in propagule suspensions of *Trichoderma viride* and *Penicillium chrysogenum*.[27] Applying *T. viride* directly (in glycerol) to wounds of 40 year old beach trees significantly reduced the level of decay (approximately 15% of the controls) due to decay fungi.[59] For the control of Dutch Elm Disease direct injection of trees with pseudomonocide has had some success.[69] Control was dependent upon the introduction of the fluorescent bacteria before the pathogen, and was also a function of the method of injection.

VI. TIMING OF DELIVERY

The timing of the application of biocontrol agents should be considered with respect to crop development, the effect of the environment upon the survival and activity of the agent and target, and other management practices (e.g., cultivation, pruning, pesticide application). Decisions on when to apply should be based not on convenience, but upon an understanding of how the above factors impinge upon the performance of the agents. Determining the proper timing of agent application could in some instances follow procedures and considerations similar to those used to determine the proper timing of fungicide application. What developmental stages of the plant are susceptible, how long it remains susceptible, and the duration of exposure to the pathogen need to be considered to determine application rates and frequencies. Also, the effects of temperature and moisture upon the survival and activity of the agent should be determined.[74,75]

With this information, appropriate decisions can be made for when and how to deliver the biocontrol agent to the environment or the plant. Coupling the introduction of an agent with irrigation or with supplemental irrigation at specific times following introduction may enhance the activity of the biocontrol agent. All these factors integrate. How many times the agent need be applied will depend not only upon plant parameters (e.g., age, duration of susceptibility, root growth and morphology, canopy architecture), but also upon agent parameters (e.g., survival, dispersal).

VII. COUPLING WITH OTHER MANAGEMENT PRACTICES

Perhaps the best chance for success of biological control will be when it is coupled to other management practices. There are many possibilities to exploit. Most biocontrol agents of soil-borne pathogens are considered to be good saprophytes and colonizers of soil. They are often associated with the recolonization of sterilized or pasteurized soils. Pasteurizing or sterilizing the soil prior to the addition of the biocontrol agent might permit a higher population of the agent to develop and perhaps result in a higher level of disease control. Fumigation of soil (Vapam, methyl bromide) with subsequent introduction of the biocontrol agent *Trichoderma harzianum*, resulted in enhanced levels of control of *Rhizoctonia solani*, *Sclerotium rolfsii*, and *Verticillium dahliae* diseases of potato.[33] Similar results were obtained for *R. solani* and *S. rolfsii* diseases of iris,[18] carrot seedlings,[76] tomatoes, and peanuts.[31] Pasteurization of soil by solar heating using a plastic mulch or tarp,[44] prior to the introduction of *T. harzianum*, resulted in significant control of *R. solani* and *V. dahliae* on potato in the field.[33] Control was greater with the combination of *T. harzianum* and solarization than with either treatment alone. Also, 90 to 100% control of *S. rolfsii* on bean in the greenhouse was achieved by combining heat treatment with the introduction of *T. harazianum*, each at levels which were ineffective when used alone.[33] To further capitalize on the benefits of solar heating of soil, a thermophilic mycoparasite (a high temperature isolate of *Paecilomyces varioti*) was introduced into soil prior to solarization for the control of diseases of peanut (Stack, J. P. and Jones, R. W., unpublished). In this case the agent was active during the solarization process. Under laboratory conditions, the *P. varioti* grew well in soil at 50°C under a wide range of soil moisture levels (-0.1 to -10.0 bars matrix potential) and readily colonized sclerotia of *Aspergillus flavus*. The colonized sclerotia failed to germinate when the temperature of the soil returned to 25 to 35°C (Stack, J. P., unpublished). Further tests with thermophilic and thermotolerant biocontrol agents in combination with solarization might be rewarding.

VIII. IMPACT ASSESSMENT

In the U.S., before biocontrol agents can be applied to the environment on a commercial scale, they must be registered with the Environmental Protection Agency and tested for potential effects upon nontarget organisms.[7,11] This includes several types of toxicity testing. The requirements for registration of conventional chemical pesticides must also be met for the registration of microbial pesticides. This will add time and cost to the development of biological control procedures. There is however, ample justification for the required (toxicity) testing. The efficacy of some biocontrol agents is linked to the production of antibiotic compounds.[24,25,67,83] Two of these compounds, chetomin and gliotoxin, have mammalian toxicity[14,19] and have been linked to diseases of sheep[14] and humans,[28,62] respectively. *Aspergillus ochraceous* has been used in field and greenhouse trials as a biocontrol agent of *Fusarium*.[57] Ochratoxin, a metabolite of *A. ochraceous*, has mammalian toxicity and is a known teratogen.[15,43,47] It has been recently determined that mycoparasitic isolates of *Penicillium* sp. and *Paecilomyces varioti* produce compounds with a suppressive effect upon the immune response of human lymphocytes.[75b]

It will not be surprising if many biocontrol agents produce secondary metabolites with wide ranging biological activities. What risk, if any, these compounds and organisms pose to public health needs to be determined. This is especially true if we intend to apply to the environment biocontrol agents with enhanced antibiotic producing potential,[24] the ability to establish close associations with the plants they are to protect,[12,56] and with resistance to the major fungicides which would be used to control them.[1,2,23] Risk assessment for biocontrol agents, whether or not genetically engineered, and the compounds they produce will be a difficult task. Such studies will need to consider exposure at all stages of biocontrol including experimental development, preparation of the agent-carrier system, application of the agent to the environment, and subsequent population and residue analyses. Scientists, support staff, industry personel, and farm workers should take exposure seriously. Risk assessment should be done with compounds singly, and in combination, to take into account synergy between toxins.[13]

IX. CONCLUDING REMARKS

The application procedures for biological control agents should be appropriate to the task to be accomplished by the agent, the presumed desirable attributes the agent must have to accomplish that task, the mechanism by which control is effected, and the factors which influence the performance of the agent. The application process should be considered and developed as a component of a total biocontrol system, not independent of the system. With this information, we can determine how best to utilize a biocontrol agent and maximize its performance. We are trying to induce epidemics in populations of plant pathogens and should therefore establish conditions conducive to development of disease in the target pathogen population.

Substrate is an important component of the application process. The activity of the agent is a function of the nature and concentration of the substrate. More research is needed in designing substrates to achieve specific agent responses. If cell wall degrading enzymes are an integral part of control then the substrate that allows for maximum and timely enzyme synthesis and excretion needs to be determined. If growth through soil is essential to effecting control, then the substrate that results in extensive hyphal development as opposed to sporulation needs to be utilized.

Prior to field application of biological control agents, public health questions need to be addressed. Considering the capabilities of some of these organisms and the compounds they produce, extreme caution should be exercised by researchers and their support staffs at all times. Perhaps through genetic engineering we may be able to design biological control agents with the desirable attributes, but from which any harmful capabilities have been excized.

REFERENCES

1. **Abd-El Moity, T. H., Papavizas, G. C., and Shatla, M. N.,** Induction of new isolates of *Trichoderma harzianum* tolerant to fungicides and their experimental use for control of white rot of onion, *Phytopathology,* 72, 396, 1982.
2. **Abd-El Moity, T. H. and Shatla, M. N.,** Biological control of white rot disease of onion (*Sclerotium cepivorum*) by *Trichoderma harzianum, Phytopathol. Z.,* 100, 29, 1981.
3. **Adams, P. B., Marios, J. J., and Ayers, W. A.,** Population dynamics of the mycoparasite, *Sporidesmium sclerotivorum,* and its host, *Sclerotinia minor,* in soil, *Soil Biol. Biochem.,* 16, 607, 1984.
4. **Aharonowitz, Y.,** Nitrogen metabolite regulation of antibiotic biosynthesis, *Annu. Rev. Microbiol.,* 34, 209, 1980.
5. **Ahmed, A. H. M. and Tribe, H. T.,** Biological control of white rot of onion (*Sclerotium cepivorum*) by *Coniothyrium minitans, Plant Pathol.,* 26, 75, 1977.
6. **Andrews, J.,** Strategies for selecting antagonistic microorganisms from the phylloplane, in *Biological Control on the Phylloplane,* Windels, G. E. and Lindow, S. E., Eds., American Phytopathological Society, St. Paul, Minn., 1985, 31.
7. **Anonymous,** Proposal for a coordinated framework for regulation of biotechnology; Notice. Part II. Office of Science and Technology Policy, *Fed. Regist.,* 49(252), 50855, 1984.
8. **Ayers, W. A. and Adams, P. B.,** Mycoparasitism and its application to biological control of plant diseases, in *Biological Control in Crop Production,* Beltsville Symposium in Agricultural Research 5, Papavizas, G. C., Ed., Allenheld, Totowa, N. J., 1981, 91.
9. **Ayers, W. A. and Adams, P. B.,** Mycoparasitism of sclerotia of *Sclerotinia* and *Sclerotium* species by *Sporidesmium sclerotivorum, Can. J. Microbiol.,* 25, 17, 1979.
10. **Backman, P. A. and Rodriguez Kabana, R.,** A system for the growth and delivery of biological control agents to the soil, *Phytopathology,* 65, 819, 1975.
11. **Battenfield, S. L.,** *Proceedings of the National Interdisciplinary Biological Control Conference,* Susan L. Battenfield, Ed., U.S. Department of Agriculture, Washington, D.C., 1983, 107.
12. **Beagle-Ristaino, J. E. and Papvizas, G. C.,** Survival and proliferation of propagules of *Trichoderma* spp. and *Gliocladium virens* in soil and in plant rhizospheres, *Phytopathology,* 75, 729, 1985.
13. **Berenbaum, M. C.,** Consequences of synergy between environmental carcinogens, *Environ. Res.,* 38, 310, 1985.
14. **Brewer, D., Duncan, J. M., Jerram, W. A., Leach, C. K., Safe, S., Taylor, A., Vining, L. C., Archibald, R. McG., Stevenson, R. G., Mirocha, C. J., and Christensen, C. M.,** Ovine ill-thrift in Nova Scotia. V. The production and toxicology of chetomin, a metabolite of *Chaetomium* spp., *Can. J. Microbiol.,* 18, 1129, 1972.
15. **Brown, M. H., Szczech, G. M., and Purmalis, B. P.,** Teratogenic and toxic effects of ochratoxin A in rats, *Toxicol. Appl. Pharmacol.,* 37, 331, 1976.
16. **Bruehl, G. W.,** Systems and mechanisms of residue possession by pioneer fungal colonists, in *Biology and Control of Soil-Borne Plant Pathogens,* Bruehl, G. W., Ed., American Phytopathological Society, St. Paul, Minn., 1975, 77.
17. **Cantwell, G. E., Cantelo, W. W., and Schroder, R. F. W.,** The integration of a bacterium and parasites to control the Colorado potato beetle and the Mexican bean beetle, *J. Entomol. Sci.,* 20, 98, 1985.

18. **Chet, I., Elad, Y., Kalfon, A., Hadar, Y., and Katan, J.,** Integrated control of soilborne and bulbborne pathogens in Iris, *Phytoparasitica,* 10, 229, 1982.
19. **Cole, R. J. and Cox, R. H.,** *Handbook of Toxic Fungal Metabolites,* Academic Press, New York, 1981, 571.
20. **Cook, R. J. and Baker, K. F.,** Introduction of antagonists for biological control, in *The Nature and Practice of Biological Control of Plant Pathogens,* American Phytopathological Society, St. Paul, Minn., 1983, 282.
21. **Cook, R. J.,** Antagonism and biological control: concluding remarks, in *Soil-Borne Plant Pathogens,* Schippers, B. and Gams, W., Eds., Academic Press, London, 1979, 653.
22. **Colyer, P. D. and Mount, M. S.,** Bacterizaiton of potatoes with *Pseudomonas putida* and its influence on postharvest soft rot diseases, *Plant Dis.,* 68, 703, 1984.
23. **Cullen, D. and Andrews, J. H.,** Benomyl-marked population of *Chaetomium globosum:* survival on apple leaves with and without benomyl and antagonism to the apple scab pathogen, *Venturia inaequalis, Can. J. Microbiol.,* 31, 251, 1985.
24. **Cullen, D. and Andrews, J. H.,** Epiphytic microbes as biological control agents, in *Plant Microbes as Biological Control Agents,* Kosuge, T. and Nester, E. W., Eds., Macmillan, New York, 1984, 381.
25. **Cullen, D. and Andrews, J. H.,** Evidence for the role of antibiosis in the antagonism of *Chaetomium globosum* to the apple scab pathogen, *Venturia inaequalis, Can. J. Bot.,* 62, 1819, 1984.
26. **Cullen, D., Berbee, F. M., and Andrews, J. H.,** *Chaetomium globosum* antagonizes the apple scab pathogen, *Venturia inaequalis,* under field conditions, *Can. J. Bot.,* 62, 1814, 1984.
27. **Dutta, B. K.,** Studies on some fungi isolated from the rhizosphere of tomato plants and the consequent prospect for the control of Verticillium wilt, *Plant Soil,* 63, 209, 1981.
28. **Eichner, R. D. and Mullbacher, A.,** Hypothesis: fungal toxins are involved in aspergillosis and AIDS, *Aust. J. Exp. Biol. Med. Sci.,* 62, 479, 1984.
29. **Elad, Y., Barak, R., and Chet, J.,** Parasitism of sclerotia of *Sclerotium rolfsii* by *Trichoderma harzianum, Soil Biol. Biochem.,* 16, 381, 1984.
30. **Elad, Y., Chet, I., and Henis, Y.,** Biological control of *Rhizoctonia solani* in strawberry fields by *Trichoderma harzianum, Plant Soil,* 60, 245, 1981.
31. **Elad, Y., Hadar, Y., Chet, I., and Henis, Y.,** Prevention, with *Trichoderma harzianum* Rifai aggr., of reinfestation by *Sclerotium rolfsii* Sacc. and *Rhizoctonia solani* Kuhn of soil fumigated with methyl bromide, and improvement of disease control in tomatoes and peanuts, *Crop Prot.,* 1, 199, 1982.
32. **Elad, Y., Kalfon, A., and Chet, I.,** Control of *Rhizoctonia solani* in cotton by seed-coating with *Trichoderma* spp. spores, *Plant Soil,* 66, 279, 1982.
33. **Elad, Y., Katan, J., and Chet, I.,** Physical, biological, and chemical control integrated for soilborne diseases in potatoes, *Phytopathology,* 70, 418, 1980.
34. **Fravel, D. R., Marois, J. J., Dunn, M. T., and Papavizas, G. C.,** Compatibility of *Talaromyces flavus* with potato seedpiece fungicides, *Soil Biol. Biochem.,* 17, 163, 1985.
35. **Fravel, D. R., Marois, J. J, Lumsden, R. D., and Connick, W. J., Jr.,** Encapsulation of potential biocontrol agents in an alginate-clay matrix, *Phytopathology,* 75, 774, 1985.
36. **Fravel, D. R. and Spurr, H. W., Jr.,** Biocontrol of tobacco brownspot disease by *Bacillus cereus* subsp. *mycoides* in a controlled environment, *Phytopathology,* 67, 930, 1977.
37. **Gindrat, D.,** Biocontrol of plant diseases by inoculation of fresh wounds, seeds, and soil with antagonists, in *Soil-Borne Plant Pathogens,* Schippers, B. and Gams, W., Eds., Academic Press, London, 1979, 541.
38. **Harder, W. and Dijkhuizen, L.,** Physiological responses to nutrient limitation, *Annu. Rev. Microbiol.,* 37, 1, 1983.
39. **Henis, Y., Lewis, J. A., and Papavizas, G. C.,** Interactions between *Sclerotium rolfsii* and *Trichoderma* spp.: relationship between antagonism and disease control, *Soil Biol. Biochem.,* 16, 391, 1984.
40. **Huber, D. M. and Watson, R. D.,** Nitrogen form and plant disease, *Annu. Rev. Phytopathol.,* 12, 139, 1974.
41. **Jager, G. and Velvis, H.,** Biological control of *Rhizoctonia solani* on potatoes by antagonists. IV. Inoculation of seed tubers with *Verticillium biguttatum* and other antagonists in field experiments, *Neth. J. Plant Pathol.,* 91, 49, 1985.
42. **Jones, R. W., Pettit, R. E., and Taber, R. A.,** Lignite and stillage: carrier and substrate for application of fungal biocontrol agents to soil, *Phytopathology,* 74, 1167, 1984.
43. **Kanisawa, M., Suzuki, S., Kozuka, Y., and Yamazcki, M.,** Histopathological studies on the toxicity of ochratoxin A in rats. I. Acute oral toxicity, *Toxicol. Appl. Pharmacol.,* 42, 55, 1977.
44. **Katan, J.,** Solar pasteurization of soils for disease control: status and prospects, *Plant Dis.,* 64, 450, 1980.
45. **Kenerley, C. M. and Stack, J. P.,** Influence of assessment methods on selection of fungal antagonists of the sclerotium-forming fungus *Phymatotrichum omnivorum, Can. J. Microbiol.,* 33, 632, 1987.
46. **Kommedahl, T. and Windels, C. E.,** Introduction of microbial antagonists to specific courts of infection: seeds, seedlings, and wounds, in *Biological Control in Crop Production,* Beltsville Symposium in Agricultural Research 5, Papavizas, G. C., Ed., Allenheld, Totowa, N.J., 1981, 227.

47. **Kubena, L. F., Phillips, T. D., Creger, C. R., Witzel, D. A., and Heidelbaugh, N. D.,** Toxicity of ochratoxin A and tannic acid to growing chicks, *Poult. Sci.,* 62, 1786, 1983.
48. **Kuter, G. A., Nelson, E. B., Hoitink, H. A. J., and Madden, L. V.,** Fungal populations in container media amended with composted hardwood bark suppressive and conducive to Rhizoctonia damping-off, *Phytopathology,* 73, 1450, 1983.
49. **Leben, C.,** Introductory remarks: biological control strategies in the phylloplane, in *Biological Control on the Phylloplane,* Windels, C. E. and Lindow, S. E., Eds., American Phytopathological Society, St. Paul, Minn., 1985, 1.
50. **Lewis, J. A. and Papavizas, G. C.,** Characteristics of alginate pellets formulated with *Trichoderma* and *Gliocladium* and their effect on the proliferation of the fungi in soil, *Plant Pathol.,* 34, 571, 1984.
51. **Lewis, J. A. and Papavizas, G. C.,** Chlamydospore formation by *Trichoderma* spp. in natural substrates, *Can. J. Microbiol.,* 30, 1, 1984.
52. **Lewis, J. A. and Papavizas, G. C.,** Production of chlamydospores and conidia by *Trichoderma* spp. in liquid and solid growth media, *Soil Biol. Biochem.,* 15, 351, 1983.
53. **Lewis, J. A. and Papavizas, G. C.,** Integrated control of Rhizoctonia fruit rot of cucumber, *Phytopathology,* 70, 85, 1980.
54. **Lockwood, J. L. and Filinow, A. B.,** Responses of fungi to nutrient-limiting conditions and to inhibitory substances in natural habitats, *Adv. Microb. Ecol.,* 5, 1, 1981.
55. **Malik, V. S.,** Genetics and biochemistry of secondary metabolism, *Adv. Appl. Microbiol.,* 28, 27, 1982.
56. **Marois, J. J., Fravel, D. R., and Papavizas, G. C.,** Ability of *Talaromyces flavus* to occupy the rhizosphere and its interactions with *Verticillium dahliae, Soil Biol. Biochem.,* 16, 387, 1984.
57. **Marois, J. J., Mitchell, D. J., and Sonada, R. M.,** Biological control of Fusarium crown rot of tomato under field conditions, *Phytopathology,* 71, 1257, 1981.
58. **Martin, S. B., Hoch, H. C., and Abawi, G. S.,** Population dynamics of *Laetisaria arvalis* and low-temperature *Pythium* spp. in untreated and pasteurized beet field soils, *Phytopathology,* 73, 1445, 1983.
59. **Mercer, P. C. and Kirk, S. A.,** Biological treatments for the control of decay in tree wounds. II. Field tests, *Ann. Appl. Biol.,* 104, 221, 1984.
60. **Mihuta, L. J. and Rowe, R. C.,** Potential biological control for fungus disease of radish, *Ohio Rep.,* 70, 9, 1985.
61. **Mount, M. S., Berman, P. M., Mortlock, R. P., and Hubbard, J. P.,** Regulation of endopolygalacturonate transeliminase in an adenosine 3′,5′-cyclic monophosphate-deficient mutant of *Erwinia carotovora, Phytopathology,* 69, 117, 1979.
62. **Mullbacher, A. and Eichner, R. D.,** Immunosuppression *in vitro* by a metabolite of a human pathogenic fungus, *Proc. Natl. Acad. Sci. USA,* 81, 3835, 1984.
63. **Nagato, N., Okumura, Y., Okamoto, R., and Ishikura, T.,** Carbon catabolite regulation of neoviridogrisein production by glucose, *Agric. Biol. Chem.,* 48, 3041, 1984.
64. **Nelson, E. B., Kuter, G. A., and Hoitink, H. A. J.,** Effects of fungal antagonists and compost age on suppression of Rhizoctonia damping-off in container media amended with composted hardwood bark, *Phytopathology,* 73, 1457, 1983.
65. **Papavizas, G. C. and Lewis, J. A.,** Introduction and augmentation of microbial antagonists for the control of soil-borne plant pathogens, in *Biological Control in Crop Producton,* Beltsville Symposium in Agricultural Research 5, Papavizas, G. C., Ed., Allenheld, Totowa, N.J., 1981, 305.
66. **Rosales, A. M. and Mew, T. W.,** Antagonistic effects of soil microorganisms on rice sheath blight pathogen, *Int. Rice Res. Newsl.,* 7, 12, 1982.
67. **Rothrock, C. S. and Gotleib, D.,** Role of antibiosis in antagonism of *Streptomyces hygroscopicus* var. *geldanus* to *Rhizoctonia solani* in soil, *Can. J. Microbiol.,* 30, 1440, 1984.
68. **Ruppel, E. G., Baker, R., Harman, G. E., Hubbard, J. P., Hecker, R. J., and Chet, I.,** Field tests of *Trichoderma harzianum* Rifai aggr. as a biocontrol agent of seedling disease in several crops and Rhizoctonia root rot of sugar beet, *Crop Prot.,* 2, 399, 1983.
69. **Scheffer, R. J.,** Biological control of Dutch Elm Disease by *Pseudomonas* species, *Ann. Appl. Biol.,* 103, 21, 1983.
70. **Sivan, A., Elad, Y., and Chet, I.,** Application of *Trichoderma harzianum* as a biocontrol agent for damping-off in vegetables, *Phytoparasitica,* 11, 3, 1983.
71. **Smith, M. S., Thomas, G. W., White, R. E., and Ritonga, D.,** Transport of *Escherichia coli* through intact and disturbed soil columns, *J. Environ. Qual.,* 14, 87, 1985.
72. **Sorensen, A. A. and Falcon, L. A.,** Microdroplet application of *Bacillus thuringiensis:* methods to increase coverage on field crops, *J. Econ. Entomol.,* 73, 252, 1980.
73. **Spurr, H. W.,** Introduction of microbial antagonists for the control of foliar plant pathogens, in *Biological Control in Crop Production,* Beltsville Symposium in Agricultural Research 5, Papavizas, G. C., Ed., Allenheld, Totowa, N.J., 1981, 323.

74. **Spurr, H. W. and Knudsen, G. R.,** Biological control of leaf diseases with bacteria, in *Biological Control on the Phylloplane,* Windels, C. E. and Lindow, S. E., Eds., American Phytopathological Society, St. Paul, Minn., 1985, 45.

75a. **Stack, J. P., Kenerley, C. M., and Pettit, R. E.,** Influence of carbon and nitrogen sources, relative carbon and nitrogen concentrations, and soil moisture on the growth in nonsterile soil of soilborne fungal antagonists, *Can. J. Microbiol.,* 33, 626, 1987.

75b. **Stack, J. P., Shephard, E. C., McMurray, D. N., Phillips, T. D., and Pettit, R. E.,** Production of immunosuppressive compounds in peanuts by thermophilic and thermotolerant fungi, *Phytopathology,* 76, 1063, 1986.

76. **Strashnow, Y., Elad, Y., Sivan, A., and Chet, I.,** Integrated control of *Rhizoctonia solani* by methyl bromide and *Trichoderma harzianum, Plant Pathol.,* 34, 146, 1985.

77. **Sylvia, D. M. and Sinclair, W. A.,** Suppressive influence of *Laccaria laccata* on *Fusarium oxysporum* and on Douglas-fir seedlings, *Phytopathology,* 73, 384, 1983.

78. **Sylvia, D. M. and Sinclair, W. A.,** Phenolic compounds and resistance to fungal pathogens induced in primary roots of Douglas-fir seedlings by the ectomycorrhizal fungi *Laccaria laccata, Phytopathology,* 73, 390, 1983.

79. **Taylor, J. B. and Guy, E. M.,** Biological control of root-infecting Basidiomycetes by species of *Bacillus* and *Clostridium, New Phytol.,* 87, 729, 1981.

80. **Utkhede, R. S. and Rahe, J. E.,** Effects of *Bacillus subtilis* on growth and protection of onion against white rot, *Phytopathol. Z.,* 106, 199, 1983.

81. **Wilkinson, H. T., Miller, R. D., and Miller, R. L.,** Infiltration of fungal and bacterial propagules into soil, *Soil Sci. Soc. Am. J.,* 45, 1034, 1981.

82. **Wilson, C. L. and Pusey, P. L.,** Potential for biological control of postharvest plant diseases, *Plant Dis.,* 69, 375, 1985.

83. **Wright, J. M.,** The production of antibiotics in soil. III. Production of gliotoxin in wheatstraw buried in soil, *Ann. Appl. Biol.,* 44, 461, 1956.

Chapter 5

POTENTIAL FOR BIOLOGICAL CONTROL OF *ASPERGILLUS FLAVUS*
AND AFLATOXIN CONTAMINATION

David M. Wilson

TABLE OF CONTENTS

I. INTRODUCTION

Biological control of the aflatoxin producing fungi, *Aspergillus flavus* Link and *Aspergillus parasiticus* Speare, is not practical at the present time. However, the literature on the effects of these fungi on crops and insects indicates that biological control of the *A. flavus* group, *A. flavus* and *A. parasiticus*, may indeed be possible.

This review will address the use of members of the *A. flavus* group as biological control agents of insects, the importance of insects in dissemination of and infection by the *A. flavus* group, and some biological interactions of host plants and the *A. flavus* group. Work on several crops that are often contaminated by aflatoxins will be emphasized; these crops include rice, cotton, tree nuts, peanuts (groundnuts), and corn (maize). The microbial interactions of the *A. flavus* group with other microorganisms and the conditions favoring aflatoxin production will also be considered.

II. INSECT AND *A. FLAVUS* GROUP INTERACTIONS

Species of the *Aspergillus flavus* group have been recognized as fungal pathogens of insects for many years. *A. flavus* and *A. parasiticus* contamination of insects in rearing facilities sometimes can become a major problem because these fungi grow in the rearing diet and eventually parasitize the insects or produce metabolites toxic to the insects. Berisford and Tsao[8,92] described the pathogenicity of *A. parasiticus* to bagworm larvae, *Thyridopteryx ephemeraeformis*, and suggested that this insect could be controlled by spraying the bag with an *A. parasiticus* spore suspension. They compared one isolate of *A. parasiticus* with an isolate of *A. flavus*, and found the *A. parasiticus* to be pathogenic, whereas the *A. flavus* was saprophytic. Sannasi and Amirthavalli[85] suggested that *A. flavus* spores could be used as a biocontrol agent by causing a mycosis of the Indian velvet mite, *Thrombidium gigas*. Field collected *Monistria discrepans* in Australia contained up to 40% *A. flavus* infected adults.[1] Survival of *M. discrepans* was high during dry weather, but many insects died during rainy weather or under high humidity laboratory conditions. Apparently environmental conditions can affect pathogenicity of insects by the *A. flavus* group.

Experiments on ear-inhabiting corn insect pests have attempted to separate parasitism of the *A. flavus* group from toxicity of their metabolites. Wilson et al.[108] compared the effect of *A. flavus* and *A. parasiticus* on corn earworm *(Heliothis zea)* and fall armyworm (*Spodoptera frugiperda*) larvae. *A. parasiticus* was more toxic than *A. flavus* to earworm larvae if the fungus was allowed to grow in the diet. The larvae were more susceptible to these fungi than the pupae and adults. When earworm larvae were inoculated with *A. flavus* and *A. parasiticus*, and transferred to fresh diet cups daily, no significant pathogenic effects were seen. The toxic metabolites of the *A. flavus* group probably are more important to the viability of ear-inhabiting corn insects than potential parasitism by the fungus. Species of the *A. flavus* group are pathogenic or toxic to parasitic insects such as *Trichospilus pupivora*, *Bracon brevicornis*,[74] *Ahasversas advena* (foreign grain beetle), *Cryptolestes pusillus* (flat grain beetle),[79] and *Epilachna vigintioctopunctata*.[54] It is difficult to separate the toxic effects from the parasitic effects of the *A. flavus* group. Aflatoxins are toxic to *Drosophila melanogaster*,[19] corn earworm larvae, fall armyworm larvae, European corn borer larvae,[68] and silkworm larvae.[70] Aflatoxin B_1 is generally more toxic than aflatoxin G_1.[68] Other toxic metabolites may be involved in the insect toxicity of the *A. flavus* group because extracts from nonaflatoxigenic isolates are frequently toxic to insects.[1] *A. flavus* could be used as a biocontrol agent for insects in certain situations, but it could not be used on any crop subject to aflatoxin contamination.

Insects can act as vectors for species of the *A. flavus* group and can introduce these fungi into susceptible plant tissue. The insects may be primary vectors such as the lesser cornstalk borer in peanut fields or the maize weevil in corn fields.[24,59,65,78] Insect damage may also create favorable habitats for growth of *A. flavus*. Examples of insects creating favorable habitats for *A. flavus* include the pink bollworm in cotton,[4,82] weevils in chestnuts,[97] and earworm damage in corn.[30,63] In addition, mites and insects that inhabit damaged plants can carry inoculum of the *A. flavus* group into susceptible plant parts and influence eventual aflatoxin contamination.[29,58,78,90]

III. *A. FLAVUS* GROUP IN RICE

Aflatoxin contamination in rice was studied by Christensen[20] and Schroeder and Boller.[12-14,86] Christensen[20] found that rice, *Oryza sativa*, stored at 17 to 20% moisture at several temperatures between 12 and 25°C, was a favorable substrate for increasing populations of *Aspergillus glaucus*, *A. candidus*, *A. flavus*, and *Penicillium* species. *A. glaucus* and *A. candidus* were the predominant species, but *A. flavus* and the *Penicillium* spp. increased at different rates. Microbial competition was important; however, the factors involved in competition were not well defined.

Schroeder and Boller[12-14,86] studied microbial interactions by inoculating rice with *A. parasiticus* and *A. chevalieri*, and *A. parasiticus* and *A. candidus*. They also ran experiments using three *A. flavus* strains that included one aflatoxin producing wild type isolate, one pigmented wild type isolate that did not produce aflatoxin, and a nonaflatoxin producing white mutant. The mutant invaded rice slowly when compared to the wild types, but when mixtures of all isolates were used as inoculum, all isolates invaded rice and retained their identity. Aflatoxin production was highest when rice was inoculated only with the aflatoxin producing strain. Mixed inoculum resulted in lowered amounts of alfatoxin after incubation at 25°C.[86]

A. parasiticus, *A. candidus*, and *A. chevalieri* were used for rice storage competition studies in Texas.[12-14] Different rates of mixed inoculum of *A. parasiticus* and *A. candidus* were used in one study of rice stored at various humidities and temperatures. When equal or fewer propagules of *A. parasiticus* than *A. candidus* were used in the initial inoculum, *A. candidus* became dominant and little or no aflatoxin was formed after 7 days at 25 to 35°C and 85, 95, or 100% relative humidity (RH). When the ratio of *A. parasiticus* to *A. candidus* was initially 7:1 only 3 to 5% as much aflatoxin was produced when compared to *A. parasiticus* alone. *A. candidus* was apparently a very good competitor in rice.[12]

Competition studies of *A. parasiticus* and *A. chevalieri* revealed more complicated interactions. *A. parasiticus* grew better at higher temperatures, and *A. chevalieri* grew better at lower temperatures; aflatoxin accumulated at all temperatures with 90% or higher RH.[14] Rice stored at 85% RH accumulated aflatoxins when inoculated with *A. parasiticus* alone, but no toxins were formed when the rice was inoculated with a mixture of *A. parasiticus* and *A. chevalieri*.

IV. *A. FLAVUS* GROUP IN COTTON

Experiments on cotton and *A. flavus* include studies on storage and seed microflora, effects of insects on field *A. flavus* infection, *A. flavus* inoculation techniques, and time of aflatoxin formation in the field. Halloin[33] used acid-delinted seed to study seed decay by *Rhizopus arrhizus*, *A. flavus*, and *A. niger*. Seed was adjusted to 20% moisture and stored at 35°C. After 9 days at 35°C, 90 to 100% of the seed and 85 to 95% of the embryos were infected by *A. flavus* and *A. niger*. The fungus isolated most frequently from uninoculated, or *R. arrhizus* inoculated seed, was *A. flavus*. *R. arrhizus* did not persist well after 21 days of storage. Seed stored in tarped field modules in California at 12% moisture or above was decayed rapidly by *Aspergillus*, *Rhizopus*, *Penicillium*, *Alternaria*, and *Hormodendron* species.[31] One characteristic of field *A. flavus* infection is the bright greenish yellow fluorescent (BGYF) color visible under ultraviolet light. With BGYF 80 to 100% of the aflatoxin is in seed locules. Removal of BGYF seed locules is more effective in removing aflatoxin than removal of fresh or stored ginned BGYF seed.[55]

Pink bollworm-damaged bolls frequently have exit holes colonized by *A. flavus*.[4] Late irrigation of cotton can increase bollworm damage and *A. flavus* infection because late irrigation promotes growth, and late bolls are highly susceptible to insect damage and *A. flavus* colonization.[82] Immature sound bolls are generally sterile and waterproof,[3] but bollworm and stinkbug injury predisposes bolls to infection by *A. flavus, A. niger,* and the *Rhizopus* sp. All of these fungi induce premature carpel separation and lint expression which enhances the probability of aflatoxin contamination, especially in years with air temperatures above 32°C for extended periods.[4,32] In Arizona, first picked (spindle picked) cotton generally has less aflatoxin than second or third picked (ground gleaned) cotton;[80] however, there are also genetic, weather, and altitude differences that affect contamination.[16] The major advantage to spindle picked cotton is that the harvester is selective for healthy open bolls that contain less *A. flavus* and aflatoxin than tight loculed bolls. Arizona cotton is most likely to be contaminated with aflatoxin in August. The toxin amounts will be maximal 30 to 35 days after boll opening. Weather and insects are more important factors influencing aflatoxin contamination than the location of the boll on the plant.[81]

No fungal interaction studies on cotton have been published, but *A. flavus* has been used to inoculate field cotton.[52,53,56] Klich and Chmielewski[52,53] successfully inoculated, until 25 days after anthesis, (bracteal) involucral nectaries with dry *A. flavus* spores. They also successfully inoculated cotyledonary leaf scars, unopened flower buds, and stigmata of new flowers. These workers suggested that *A. flavus* infection may be systemic in cotton plants. Another inoculation technique that has been successful is mechanically air blasting *A. flavus* contaminated soil onto cotton plants.[56]

V. *A. FLAVUS* GROUP IN PECANS, ALMONDS, AND OTHER TREE NUTS AND FIGS

Wells and Payne[96] found that weevil damaged pecans frequently contained *Alternaria, Epicoccum, Penicillium, Pestalotia, Monochaeta, Cladosporium,* and *Fusarium* species, but infrequently were contaminated by the *Aspergillus* sp., especially *A. flavus*. Weevil damaged chestnuts[97] contained *Penicillium, Rhizopus, Alternaria, Aspergillus,* and *Fusarium* species. *P. citrinum* was the most frequently isolated species from weevil damaged chestnuts, but *A. wentii* and *A. flavus* also were isolated frequently. Huang and Hanlin[42] isolated 44 fungal genera and 119 species from pecans which were fresh (on the tree) and in the market place. The fresh pecans were frequently contaminated with *Penicillium, Aspergillus, Pestalotia, Rhizopus,* and *Fusarium* species; whereas, the market pecans had *Aspergillus, Penicillium, Eurotium,* and *Rhizopus* species. *A. niger* was predominant in fresh pecans, while *A. niger, A. flavus,* and the *A. glaucus* group were frequent in market pecans. Beuchat et al.[9,10] also found a low incidence of *A. flavus* in fresh pecans with a high incidence of internal mold. Processed market nuts have a higher incidence of fungi in general than fresh and processed products. Processed products generally have lower mold counts than they had before processing.[95] *A. flavus* has been found in cashew nuts, hazel nuts, pecan nuts, currants, unbleached sultanas, pears, raisins, figs, peaches, and prunes.[95] Figs are not colonized until they are ripe, and *A. flavus* and aflatoxin can be a problem in figs especially under sun drying conditions.[17]

The California studies on *A. flavus* and aflatoxin contamination of almonds are among the best microbial interaction models available for assessing the biocontrol potential of *A. flavus*.[75,76] *A. flavus* in almonds was found more frequently in sunny locations than in shady locations of orchards, and the incidence of *A. flavus* increased as the nuts dried. There was a positive correlation between mean temperature and the incidence of *A. flavus*. Warmer temperatures favored *A. flavus*. *A. flavus*, *A. niger*, *A. wentii*, *A. glaucus*, *A. ochraceus*, *Penicillium*, *Alternaria*, and *Rhizopus* species were all found in almonds, but a negative correlation was seen between the incidence of *A. flavus* and *Alternaria* and *Penicillium* species.[76] Inoculation of almonds with *A. flavus* and other fungi showed that antagonism could be a potential biological control method. *Monilinia fructicola*, *Ulocladium atrum*, *U. chartarum*, *Drechslera spicifera*, *Fusarium roseum*, *Rhizopus stolonifer*, *Cladosporium cladosporioides*, *Penicillium funiculosum*, and *Aspergillus ficuum* were used as potential antagonists. The moisture requirements of the fungi were important in antagonism of *A. flavus*. *U. chartarum* showed the most antagonistic activity toward *A. flavus*.[75]

VI. *A FLAVUS* GROUP IN PEANUTS (GROUNDNUTS)

The aflatoxin problem was first brought to the forefront because of toxicity problems associated with Brazilian peanut meal. Austwick and Ayerst[7] found that the *A. flavus* group was dominant in Brazilian and African peanut kernels and meal; *A. tamarii*, *A. niger*, and the *Phoma* sp. were also often associated with meal and kernels. In Israel,[50] *A. niger* brought into storage from the field was dominant in stored peanuts while the incidence of *Penicillium* sp. declined. *A. flavus* and the *Fusarium* sp. were infrequently isolated.

Diener and Davis[25] found that the minimum relative humidity for *A. flavus* group growth was about 85% RH, and moisture levels of 9 to 11% in kernels favored growth of *A. flavus* which could grow at temperatures from 12 to 40°C with an optimum between 25 and 35°C. *A. flavus* and *A. parasiticus* can both infect sound peanut seed and produce equivalent amounts of B_1 while only *A. parasiticus* produces G_1.[104] The substrate and the isolate can both influence the amount of aflatoxin produced in storage.

The occurrence of fungi in peanut field soil, in the rhizosphere, and in the geocarposphere has been investigated at several locations worldwide.[5,34,43,44,46,47,49,61,71,77,105] Israeli soils generally have low populations of the *A. flavus* group, but *A. flavus* occurs more often in medium and heavy soils than light soils. Joffe[47] listed 157 fungal species that occur in soil, in the peanut rhizosphere, and the peanut geocarposphere.[46,47] *A. niger*, *P. funiculosum*, and *P. rubrum* were most commonly isolated, while *A. flavus* was widespread but occurred in small quantities.[49] Other fungi often isolated include *Rhizopus oryzae*, *Fusarium oxysporum*, and *F. solani*.

Jackson[43,44] and Hanlin[34] determined the Georgia peanut microflora in the 1960s. These workers found that *P. funiculosum*, *P. rubrum*, *P. citrinum*, *A. niger*, *A. flavus*, *Fusarium* sp., *Rhizopus* sp., *Phoma* sp. and *Gliocladium* sp. were all commonly isolated. *A. flavus* was generally present in low numbers, but populations increased when pods were rehydrated.[44] Pegs had a higher incidence of *A. flavus* than mature pods. Wilson and Flowers[105] found no relationship between *A. flavus* and crop rotation practices. In India, Rao[77] found that *Aspergillus* species were dominant, but *A. flavus* was rare in the peanut rhizosphere; whereas others found that *A. flavus* was common in Nigeria,[61] especially in damaged pods. In Texas, *A. flavus* and *A. niger* were dominant in Spanish peanuts before harvest and *A. flavus* was often evident in visible openings in the pod.[71] McDonald[61] found that with normal pod microflora there was little kernel infection by fungi but kernels in damaged pods were often infected by fungi. Mites tolerate *A. flavus* well and may be vectors of *A. flavus*.[6] Lesser cornstalk borer larvae can be excellent vectors of *A. flavus*,[59] and lesser cornstalk borer damage or termite damage both provide excellent habitats for *A. flavus* growth.[24,62]

Extended drought before digging favors *A. flavus* invasion of peanuts.[24] Wilson and Stansell[109] found that irrigation throughout the season prevented aflatoxin contamination in sound kernels, and drought during the last 45 to 70 days did not always lead to aflatoxin contamination. Cole and co-workers[11,21,22,38,83,84] devised several rainfall shelters where they were able to control applied water and modify soil temperatures. They found that aflatoxin was formed in sound kernels with drought and that irrigation prevented aflatoxin contamination of sound kernels. More aflatoxin was produced in immature, damaged, and inedible kernels than in sound kernels. *A. flavus* was recovered from 56% of sound mature kernels from drought plots and from 7% of sound kernels from irrigated plots. The ratio of *A. flavus* to *A. niger* was variable. When this ratio was greater than 19 to 1 toxin contamination was frequent, and when it was below 9 to 1 aflatoxin contamination did not occur. Irrigation favored *A. niger* over *A. flavus*. The minimum mean temperature where aflatoxins were found in sound kernels was about 26°C and the maximum mean temperature was about 33°C.

The only biocontrol paper on *A. flavus* in peanuts was published by Mixon, Bell, and Wilson.[69] They applied a *Trichoderma harzianum* preparation to soil amended with various rates of gypsum. The gypsum plus *T. harzianum* treatments reduced aflatoxin over *T. harzianum* alone, but there were treatment × genotype interactions.

VII. *A. FLAVUS* GROUP IN CORN (MAIZE)

Corn grown in Zambia, or the corn belt of the U.S., rarely is colonized by high populations of the *A. flavus* group in the field,[36,60] whereas corn grown in the Southeastern U.S. frequently is heavily colonized by *A. flavus* and contaminated with aflatoxin in the field.[2,21,27,40,66,67,73] Aflatoxin contamination of corn is a field as well as a storage problem. Several recent excellent reviews on aflatoxin contamination of corn could be consulted for greater detail.[35,57,87,103] Draughon[28] recently reviewed the effects of pesticides on aflatoxin contamination of corn.

Microbial quality of corn at Midwestern mills was primarily correlated with the mills rather than with the month samples taken. *Penicillium, Fusarium, Aspergillus, Helminthosporium, Nigrospora,* and *Trichoderma* species were most frequently isolated.[15] *A. flavus* ranged from 4.5 to 6.8% and did not vary with storage time. Hill et al.[39] found *Aspergillus, Penicillium, Cladosporium, Alternaria, Helminthosporium,* and *Fusarium* species in Georgia corn dust. *A. flavus* was more prevalent in dust sampled in 1980, a drought year, than in dust collected in 1979 and 1982.

Wilson and Bell[104] found that *A. flavus* and *A. parasiticus* produced equal amounts of aflatoxin B_1 in inoculated corn with only *A. parasiticus* producing the G aflatoxins. Wicklow and Hesseltine[99] inoculated autoclaved corn with *A. flavus* along with several other fungi. BGYF was produced when *A. flavus* was inoculated with *Alternaria alternata, Cladosporium cladosporioides, Curvularia lunata, Fusarium moniliforme,* and *Penicillium variable.*[99] *T. viride* inoculation did not allow the growth of *A. flavus.*[99]

Corn silks and ear inhabiting insects were collected in Georgia, South Carolina, Texas, Missouri, Illinois, and Iowa, and assessed for the presence of *A. flavus.*[29] *A. flavus* was widespread and present on at least one silk and one insect at all locations, and the incidence of *A. flavus* associated with insects varied from 1.7 to 3.1%. Two corn hybrids were planted at eleven locations with planting and harvest dates staggered to determine if *A. flavus* was associated with ear inhabiting insects.[58] Overall 3.3% of the insects contained *A. flavus* but the frequency did not vary by location. In those tests there was more aflatoxin in Southern locations than Midwestern locations. In other tests Fennell and co-workers[30] found a correlation between corn earworm damage and aflatoxin, with a regional effect on the incidence of *A. flavus* on larvae. In Georgia, McMillian[63] found a high incidence of *A. flavus* on insects. Pesticide control of insects lowered but did not eliminate aflatoxin contamination.[63] Rodriguez[78] and McMillian et al.[65] found that maize weevils (*Sitophilus zeamis*) were good vectors of *A. flavus* and *A. parasiticus,* and that weevil damage by weevils carrying aflatoxigenic strains was followed by a great deal of aflatoxin contamination. When weevils were exposed to nontoxic strains additional aflatoxin production was not observed.[78]

Corn can be directly invaded by the *A. flavus* group without insect damage. Payne, at North Carolina State University, demonstrated that silk colonization was possible followed by direct kernel invasion.[73] Temperatures around 30°C were optimum for colonization of yellow brown silk. The kernels were invaded much later in the dent stage. *Helminthosporium maydis* ear rot and other ear rots may also predispose corn to *A. flavus* infection.[27] The benchmark experiments on aflatoxin in corn carried out by Anderson et al.[2] clearly demonstrated the importance of insect damage, bran coat damage, dense corn populations, low fertility, cultivars, and stress conditions on aflatoxin buildup. Aflatoxin contamination was especially related to bran coat integrity. McMillian et al.[64] demonstrated that heavy dews may contribute to *A. flavus* infection and elevate toxin amounts. Studies over a 6-year period in Georgia revealed positive correlations between insect damage, aflatoxin amounts, and visible *A. flavus* on corn ears, positive correlations between visible *A. flavus*, insect damage and aflatoxin contamination, and positive correlations between aflatoxin, temperature and net evaporation.[66,67] Irrigation is beneficial, but is less effective in lowering aflatoxin contamination in corn than it is in peanuts.[21] Hill et al.[40] reported positive correlations between *A. flavus* and *Penicillium* species in developing corn ears, but negative correlations between *A. flavus* and *F. moniliforme*, *Cladosporium* sp., *Alternaria* sp., *Helminthosporium* sp., bacteria, and actinomyces in developing corn ears. Temperature was the single most important factor influencing *A. flavus* and aflatoxin in developing corn ears. Wilson et al.[106] identified several fungistatic corn volatiles, including beta-ionone, that may affect colonization by *A. flavus*.

Wicklow et al.[100] inoculated corn with mixtures of *A. flavus* and *A. niger*, and *A. flavus* and *T. viride*, and found that the competition lowered the amounts of aflatoxin produced. Calvert et al.[18] inoculated corn with different ratios of *A. flavus* and *A. parasiticus*. All ratios gave equivalent amounts of B_1 but the G_1 production was not linear with inoculum rates. *A. parasiticus* apparently did not compete well with *A. flavus* in these tests. Wilson et al.[107] inoculated corn with several *A. flavus* and *A. parasiticus* mutants. The *A. parasiticus* mutant produced aflatoxin B_1, B_2, G_1, and G_2, whereas the *A. flavus* mutants produced no toxin or only B_1 and B_2. The *A. parasiticus* mutant that produced the B and G aflatoxins seemed to be more aggressive and survived better than the *A. flavus* mutants. The nontoxic *A. flavus* mutant lowered the amount of aflatoxin at harvest over the control, but when wild type aflatoxigenic *A. flavus* contaminated weevils were introduced after the nontoxic mutant became established, the maize weevil activity resulted in highly contaminated corn at harvest.

VIII. BIOLOGY OF THE *A. FLAVUS* GROUP

Aflatoxin is only produced by two species of the *A. flavus* group. Wicklow[98] considered *A. flavus* and *A. parasiticus* to be valid species, and considered *A. toxicarius* to be synonymous with *A. parasiticus*. The remaining species of the *A. flavus* group do not produce aflatoxins.

A. flavus and *A. parasiticus* both require exogenous carbon and nitrogen for germination.[72] Amino acid mixtures, especially proline and alanine or glucose plus NH_4Cl, are more favorable for germination than other single amino acids or sugars. Germination is inhibited by high spore densities, and CO_2 is required for germination. Pass and Griffin[72] suggested that injury to plant cells was important in successful *A. flavus* germination and colonization. Hora and Baker[41] found that *A. flavus* germination was inhibited by fungistatic soil volatiles. Kaminski et al.[51] identified several volatiles from *A. flavus* growing on wheat meal. The volatiles were vacuum distilled and included 3-methyl butanol, 3-octanone, 3-octanol, 1-octen-3-ol, 1-octanol, and *cis*-2-octen-1-ol. The octenols were predominant and may be useful in the development of a growth index for *A. flavus*.

Hesseltine et al.[37] tested 67 *A. flavus* group strains for the production of aflatoxin B_1, G_1, and M_1. Eleven of 14 *A. parasiticus* isolates produced B_1, G_1, and M_1. Three isolates produced low amounts of B_1. Sixteen *A. flavus* isolates produced B_1 and M_1, five produced traces of B_1, and six *A. flavus* isolates did not produce any aflatoxin. Atypical *A. flavus* isolates sometimes produced B_1, G_1, and M_1. The authors stated that the taxonomy was imperfect and the distinctions between *A. flavus* and *A. parasiticus* were not always clear cut.[37] Wicklow and Shotwell[101] found aflatoxin B_1 and G_1 in sclerotia of both *A. flavus* and *A. parasiticus*. Diener and Davis[26] found that about 80% of their *A. flavus* group isolates were toxic to some degree, and of these about 90% produced B_1 and about 10% produced B_1 and G_1. The optimum temperatures for *A. parasiticus* and *A. flavus* were between 25 and 35°C.[23] Van Walbeek et al.[93] found that *A. flavus* can produce aflatoxins slowly at 7.5 and 10°C. Aflatoxins are rarely produced below 10°C or above 40°C.[40]

IX. BIOLOGICAL CONTROL OF THE *A. FLAVUS* GROUP

A. flavus may survive by colonizing and sporulating on organic matter or root segments in the soil,[89] or viable sclerotia may germinate in or on the surface of the soil and provide conidia for wind or insect dispersal.[102] Sclerotia can be invaded by other fungi in the soil which may or may not suppress germination. *A. flavus* sclerotia may be colonized by the *Gliocladium* spp., *Paecilomyces varioti*, and the *Trichoderma* sp., any of which can render the sclerotia inviable or suppress germination.[88]

Biological control of aflatoxin and the *A. flavus* group could be based on soil fungistasis,[41] soil antagonism,[48,91] competition,[5,45,94] or microbial degradation.[5,45] Three *Streptomyces* sp. isolated from the soil inhibited germination of *A. flavus* at 25°C.[91] *R. oryzae*, in mixed culture with *A. flavus*, inhibited *A. flavus* and metabolized aflatoxin, while mixed cultures of *Scopulariopsis brevicaulis*, *Norcardia* sp., and *A. niger* all detoxified aflatoxigenic *A. flavus*.[45] Mixed cultures of *A. parasiticus* with *R. nigricans*, *Saccharomyces cerevisiae*, *Acetobacter aceti*, and *Brevibacterium linens* showed either inhibition or stimulation of aflatoxins. *R. nigricans* and *S. cerevisiae* inhibited *A. parasiticus* growth and toxin production.[94] Weckbach and Morth[94] suggested that *R. nigricans* may metabolize aflatoxins. *B. linens* showed a slight inhibition, while *A. aceti* stimulated *A. parasiticus* growth and toxin production.[94] *A. niger*, *F. roseum*, and *Macrophomina phaseoli* may affect aflatoxin production slightly in peanuts.[5] Joffe[48] reported a slight antagonism between *A. flavus* and several fungi including *A. niger*, *P. funiculosum*, *P. rubrum*, or *F. solani* in peanuts. This antagonism apparently exists in nature, but the fungi coexist very well in peanuts.

X. CONCLUSION

The aflatoxin problem in crop plants is primarily a quality problem. In the broad sense biological control measures are presently being used to control the growth of *A. flavus* and *A. parasiticus*, and aflatoxin production in several foods and feeds. Environmental, cultural, and nutrient manipulations such as irrigation, nitrogen and calcium nutrition, plant density, insect control, and good management can all be used to help lower aflatoxin contamination. There is good evidence that successful biological control strategies for the *A. flavus* group can be devised.

REFERENCES

1. **Allsopp, P. G.**, Seasonal history, hosts and natural enemies of *Monistria discrepans* (Walker) (Orthoptera: Pyrgomorphidae) in South-West Queensland, *J. Aust. Entom. Soc.*, 17, 65, 1978.
2. **Anderson, H. W., Nehring, E. W., and Wichser, W. R.**, Aflatoxin contamination of corn in the field, *J. Agric. Food Chem.*, 23, 775, 1975.
3. **Ashworth, L. J., Jr. and Hine, R. B.**, Structural integrity of the cotton fruit and infection by microorganisms, *Phytopathology*, 61, 1245, 1971.
4. **Ashworth, L. J., Jr., Rice, R. E., McMeans, J. L., and Brown, C. M.**, The relationship of insects to infection of cotton bolls by *Aspergillus flavus*, *Phytopathology*, 61, 488, 1971.
5. **Ashworth, L. J., Jr., Schroeder, H. W., and Langley, B. C.**, Aflatoxins: environmental factors governing occurrence in Spanish peanuts, *Science*, 148, 1228, 1965.
6. **Aucamp, J. L.**, The role of mite vectors in the development of aflatoxin in groundnuts, *J. Stored Prod. Res.*, 5, 245, 1969.
7. **Austwick, P. K. C. and Ayerst, G.**, Groundnut microflora and toxicity, *Chem. Ind.*, Jan., 55, 1963.
8. **Berisford, Y. C. and Tsao, C. H.**, Distribution and pathogenicity of fungi associated with the bagworm, *Thyridopteryx ephemeraeformis* (Hayworth), *Environ. Entomol.*, 4, 257, 1975.
9. **Beuchat, L. R.**, Incidence of molds on pecan nuts at different points during harvesting, *Appl. Microbiol.*, 29, 852, 1975.
10. **Beuchat, L. R., Williamson, R. E., and Holmes, M. R.**, Preliminary studies on the incidence of molds on pecan nuts during harvesting and drying, *Pecan South*, 2, 244, 1975.
11. **Blankenship, P. D., Cole, R. J., Sanders, T. H., and Hill, R. A.**, Effect of geocarposphere temperature on pre-harvest colonization of drought-stressed peanuts by *Aspergillus flavus* and subsequent aflatoxin contamination, *Mycopathologia*, 85, 69, 1984.
12. **Boller, R. A. and Schroeder, H. W.**, Influence of *Aspergillus candidus* on production of aflatoxin in rice by *Aspergillus parasiticus*, *Phytopathology*, 64, 121, 1974.
13. **Boller, R. A. and Schroeder, H. W.**, Influence of *Aspergillus chevalieri* on production of aflatoxin in rice by *Aspergillus parasiticus*, *Phytopathology*, 63, 1507, 1973.
14. **Boller, R. A. and Schroeder, H. W.**, Accumulation of aflatoxins in stored rice in relation to competition between *Aspergillus parasiticus and A. chevalieri*, *Phytopathology*, 58, 725, 1968.
15. **Bothast, R. J., Rogers, R. F., and Hesseltine, C. W.**, Microbial survey in corn in 1970-71, *Cereal Sci. Today*, 18, 22, 1973.
16. **Brown, C. M., Ashworth, L. J., Jr., and McMean, J. L.**, Differential response of cotton varieties to infection by *Aspergillus flavus*, *Crop Sci.*, 15, 276, 1975.
17. **Buchanan, J. R., Sommer, N. F., and Fortlage, R. J.**, *Aspergillus flavus* infection and aflatoxin production in fig fruits, *Appl. Microbiol.*, 30, 238, 1975.
18. **Calvert, O. H., Lillehoj, E. B., Kwolek, W. F., and Zuber, M. S.**, Aflatoxin B_1 and G_1 production in developing *Zea mays* kernels from mixed inocula of *Aspergillus flavus* and *A. parasiticus*, *Phytopathology*, 68, 501, 1978.
19. **Chinnici, J. P., Booker, M. A., and Llewellyn, G. C.**, Effect of aflatoxin B_1 on viability, growth, fertility, and crossing over in *Drosophila melanogaster* (Diptera), *J. Invertebr. Pathol.*, 27, 255, 1976.
20. **Christensen, C. M.**, Influence of moisture content, temperature, and time of storage upon invasion of rough rice by storage fungi, *Phytopathology*, 59, 145, 1969.
21. **Cole, R. J., Hill, R. A., Blankenship, P. D., Sanders, T. H. and Garren, K. H.**, Influence of irrigation and drought stress on invasion by *Aspergillus flavus* of corn kernels and peanut pods, *Dev. Ind. Microbiol.*, 23, 229, 1982.
22. **Cole, R. J., Sanders, T. H., Hill, R. A., and Blankenship, P. D.**, Mean geocarposphere temperatures that induce preharvest aflatoxin contamination of peanuts under drought stress, *Mycopathologia*, 91, 41, 1985.
23. **Davis, N. D. and Diener, U. L.**, Some characteristics of toxigenic and nontoxigenic isolates, in *Aflatoxin and Aspergillus flavus in Corn*, Diener, U. L., Asquith, R. L., and Dickens, J. W., Eds., Southern Cooperative Series Bulletin 279 for Southern Regional Research Project S-132, Auburn, Ala., 1983, 1.
24. **Dickens, J. W., Satterwhite, J. B., and Sneed, R. E.**, Aflatoxin-contaminated peanuts produced on North Carolina farms in 1968, *Proc. Am. Peanut Res. Educ. Assoc.*, 5, 48, 1973.
25. **Diener, U. L. and Davis, N. D.**, Aflatoxin formation in peanuts by *Aspergillus flavus*, *Ala. Agric. Exp. Stn. Bull.*, 493, 49, 1977.
26. **Diener, U. L. and Davis, N. D.**, Aflatoxin production by isolates of *Aspergillus flavus*, *Phytopathology*, 56, 1390, 1966.
27. **Doupnik, B., Jr.**, Maize seed predisposed to fungal invasion and aflatoxin contamination by *Helminthosporium maydis* ear rot, *Phytopathology*, 62, 1367, 1972.

28. **Draughon, F. A.,** Control or suppression of aflatoxin production with pesticides, in *Aflatoxin and Aspergillus flavus in Corn,* Diener, U. L., Asquith, R. L., and Dickens, J. W., Eds., Southern Cooperative Series Bulletin 279 for Southern Regional Research Project S-132, Auburn, Ala., 1983, 81.

29. **Fennell, D. I., Kwolek, W. F., Lillehoj, E. B., Adams, G. L., Bothast, R. J., Zuber, M. S., Calvert, O. H., Guthrie, W. D., Bockholt, A. J., Manwiller, A., and Jellum, M. D.,** *Aspergillus flavus* presence in silks and insects from developing and mature corn ears, *Cereal Chem.,* 54, 770, 1977.

30. **Fennell, D. I., Lillehoj, E. B., Kwolek, W. F., Guthrie, W. D., Sheely, R., Sparks, A. N., Widstrom, N. W., and Adams, G. L.,** Insect larval activity on developing corn ears and subsequent aflatoxin contamination of seed, *J. Econ. Entomol.,* 71, 624, 1978.

31. **Garber, R. H., Ferguson, D., and Carter, L. M.,** Internal populations of fungi and bacterial in cottonseed stored under various environmental conditions, *Phytopathology,* 73, 958, 1983.

32. **Gardner, D. E., McMeans, J. L., Brown, C. M., Bilbrey, R. M., and Parker, L. L.,** Geographical localization and lint fluorescence in relation to aflatoxin production in *Aspergillus flavus* — infected cottonseed, *Phytopathology,* 64, 452, 1974.

33. **Halloin, J. M.,** Postharvest infection of cottonseed by *Rhizopus arrhizus, Aspergillus niger,* and *Aspergillus flavus, Phytopathology,* 65, 1229, 1975.

34. **Hanlin, R. T.,** Fungi in developing peanut fruits, *Mycopathologia,* 38, 93, 1969.

35. **Hesseltine, C. W.,** Conditions leading to mycotoxin contamination of foods and feeds, in *Mycotoxins and Other Fungal Related Food Problems,* Rodricks, J. V., Ed., Advances in Chemistry Series No. 149, American Chemical Society, Washington, D.C., 1976, 1.

36. **Hesseltine, C. W. and Bothast, R. J.,** Mold development in ears of corn from tasseling to harvest, *Mycologia,* 69, 328, 1977.

37. **Hesseltine, C. W., Shotwell, O. L., Smith, M., Ellis, J. J., Vandegraft, E., and Shannon, G.,** Production of various aflatoxins by strains of the *Aspergillus flavus* series, 1st U. S. — Japan Conf. on Toxic Micro-Organisms, U.J.N.R. Joint Panels on Toxic Microorganisms, U.S. Department of the Interior, Washington, D.C., 1970, 202.

38. **Hill, R. A., Blankenship, P. D., Cole, R. J., and Sanders, T. H.,** Effects of soil moisture and temperature and preharvest invasion of peanuts by the *Aspergillus flavus* group and subsequent aflatoxin development, *Appl. Environ. Microbiol.,* 45, 628, 1983.

39. **Hill, R. A., Wilson, D. M., Burg, W. R., and Shotwell, O. L.,** Viable fungi in corn dust, *Appl. Environ. Microbiol.,* 47, 84, 1984.

40. **Hill, R. A., Wilson, D. M., McMillian, W. W., Widstrom, N. W., Cole, R. J., Sanders, T. H., and Blankenship, P. D.,** Ecology of the *Aspergillus flavus* group and aflatoxin formation in maize and groundnut, in *Trichothecenes and Other Mycotoxins,* Lacey, J., Ed., John Wiley & Sons, New York, 1985, 79.

41. **Hora, T. S. and Baker, R.,** Volatile factor in soil fungistasis, *Nature,* 225, 1071, 1970.

42. **Huang, L. H. and Hanlin, R. T.,** Fungi occurring in freshly harvested and in-market pecans, *Mycologia,* 67, 689, 1975.

43. **Jackson, C. R.,** A field study of fungal associations on peanut fruit, *Res. Bull. Univ. Ga. Coll. Ag. Exp. Stn.,* 26, 29, 1968.

44. **Jackson, C. R.,** Peanut pod mycoflora and kernel infection, *Plant Soil,* 23, 203, 1965.

45. **Jarvis, B.,** Factors affecting the production of mycotoxins, *J. Appl. Bacteriol.,* 34, 199, 1971.

46. **Joffe, A. Z.,** The mycoflora of fresh and stored groundnut kernels in Israel, *Mycopathologia,* 39, 256, 1969.

47. **Joffe, A. Z.,** The mycoflora of groundnut rhizosphere, soil and geocarposphere on light, medium and heavy soils and its relations to *Aspergillus flavus, Mycopathologia,* 37, 150, 1969.

48. **Joffe, A. Z.,** Relationships between *Aspergillus flavus, A. niger,* and some other fungi in the mycoflora of groundnut kernels, *Plant Soil,* 31, 57, 1969.

49. **Joffe, A. Z. and Borut, S. Y.,** Soil and kernel mycoflora of groundnut fields in Israel, *Mycologia,* 58, 629, 1966.

50. **Joffe, A. Z. and Lisker, N.,** The mycoflora of fresh and subsequently stored groundnut kernels on various soil types, *Israel J. Bot.,* 18, 77, 1969.

51. **Kaminski, E., Libbey, L. M., Stawicki, S., and Wasowicz, E.,** Identification of the predominant volatile compounds produced by *Aspergillus flavus, Appl. Microbiol.,* 24, 721, 1972.

52. **Klich, M. A.,** Field studies on the mode of entry of *Aspergillus flavus* into cotton seeds, *Mycologia,* 76, 665, 1984.

53. **Klich, M. A. and Chmielewski, M. A.,** Nectaries as entry sites for *Aspergillus flavus* in developing cotton bolls, *Appl. Environ. Microbiol.,* 50, 602, 1985.

54. **Krishnamoorthy, C. and Naidu, M. S.,** *Aspergillus flavus* Link — A fungal parasite of *Epilachna vigintioctopunctata* F., *Curr. Sci.,* 40, 666, 1971.

55. **Lee, L. S., Cucullu, A. F., Pons, W. A., Jr., and Russell, T. E.,** Separation of aflatoxin-contaminated cottonseed based on physical characteristics of seed cotton and ginned seed, *J. Am. Oil Chem. Soc.,* 54, 238A, 1977.

56. **Lee, L. S., Lee, L. V., Jr., and Russell, T. E.,** Aflatoxin in Arizona cottonseed: field inoculation of bolls by *Aspergillus flavus* spores in wind-driven soil, *J. Am. Oil Chem. Soc.,* 63, 530, 1986.

57. **Lillehoj, E. B.,** Effect of environmental and cultural factors on aflatoxin contamination of developing corn kernels, in *Aflatoxin and Aspergillus flavus in Corn,* Diener, U. L., Asquith, R. L., and Dickens, J. W., Eds., Southern Cooperative Series Bulletin 279 for Southern Regional Research Project S-132, Auburn, Ala., 1983, 27.

58. **Lillehoj, E. B., Fennell, D. I., Kwolek, W. F., Adams, G. L., Zuber, M. S., Horner, E. S., Widstrom, N. W., Warren, H., Guthrie, W. D., Sauer, D. B., Findley, W. R., Manwiller, A., Josephson, L. M., and Bockholt, A. J.,** Aflatoxin contamination of corn before harvest: *Aspergillus flavus* association with insects collected from developing ears, *Crop Sci.,* 18, 921, 1978.

59. **Lynch, R. E. and Wilson, D. M.,** Relation of lesser cornstalk borer damage to peanut pods and the incidence of *Aspergillus flavus, Proc. Am. Peanut Res. Educ. Soc.,* 16, 35, 1984.

60. **Marasas, W. F. O., Kriek, N. P. J., Steyn, M., Rensburg, S. J., and Van Schalkwyk, D. J.,** Mycotoxicological investigations of Zambian maize, *Food. Cosmet. Toxicol.,* 16, 39, 1978.

61. **McDonald, D.,** Soil fungi and the fruit of the groundnut *(Arachis hypogaea* L.), Samaru Miscellaneous Paper No. 28, Institute Agricultural Research, Ahmadu Bello University, Samaru, Zaria, Nigeria, 1968, 31.

62. **McDonald, D. and Harkness, C.,** Growth of *Aspergillus flavus* production of aflatoxin in groundnuts, *Trop. Sci. Pt. II,* 5, 208, 1963.

63. **McMillian, W. W.,** Role of arthropods in field contamination, in *Aflatoxin and Aspergillus flavus in Corn,* Diener, U. L., Asquith, R. L., and Dickens, J. W., Eds., Southern Cooperative Series Bulletin 279 for Southern Regional Research Project S-132, Auburn, Ala., 1983, 20.

64. **McMillian, W. W., Widstrom, N. W., and Wilson, D. M.,** Insect damage and aflatoxin contamination in preharvest corn: influence of genotype and ear wetting, *J. Entomol. Sci.,* 20, 66, 1985.

65. **McMillian, W. W., Widstrom, N. W., Wilson, D. M., and Hill, R. A.,** Transmission by maize weevils of *Aspergillus flavus* and its survival on selected corn hybrids, *J. Econ. Entomol.,* 73, 793, 1980.

66. **McMillian, W. W., Wilson, D. M., and Widstrom, N. W.,** Aflatoxin contamination ιₓ ρᵣcharvest corn in Georgia: a six-year study of insect damage and visible *Aspergillus flavus, J. Environ. Qual.,* 14, 200, 1985.

67. **McMillian, W. W., Wilson, D. M., and Widstrom, N. W.,** Insect damage, *Aspergillus flavus* ear mold, and aflatoxin contamination in South Georgia corn fields in 1977, *J. Environ. Qual.,* 7, 564, 1978.

68. **McMillian, W. W., Wilson, D. M., Widstrom, N. W., and Perkins, W. D.,** Effects of aflatoxin B_1 and G_1 on three insect pests of maize, *J. Econ. Entomol.,* 73, 26, 1980.

69. **Mixon, A. C., Bell, D. K., and Wilson, D. M.,** Effects of chemical and biological agents on the incidence of *Aspergillus flavus* and aflatoxin contamination of peanut seed, *Phytopathology,* 74, 1440, 1984.

70. **Mura Koshi, S., Ohtomo, T., and Kurata, H.,** Toxic effects of various mycotoxins to silkworm *(Bombyx mori* L.) larvae in ad libitum feeding test, *J. Food Hyg. Soc., Japan,* 14, 65, 1973.

71. **Norton, D. C., Menon, S. K., and Flangas, A. L.,** Fungi associated with unblemished Spanish peanuts in Texas, *Plant Dis. Rep.,* 40, 374, 1956.

72. **Pass, T. and Griffin, G. J.,** Exogenous carbon and nitrogen requirements for conidial germination by *Aspergillus flavus, Can. J. Microbiol.,* 18, 1453, 1972.

73. **Payne, G. A.,** Nature of field infection of corn by *Aspergillus flavus,* in *Aflatoxin and Aspergillus flavus in Corn.* Diener, U. L., Asquith, R. L., and Dickens, J. W., Eds., Southern Cooperative Series Bulletin 270 for Southern Regional Research Project S-132, Auburn, Ala., 1983, 16.

74. **Peethambaran, C. K., Wilson, K. I., and Panicker, K. G. K.,** *Aspergillus flavus* Link — Parasitic in *Trichospilus pupivora* Ferr and *Bracon brevicornis* Wesmael, *Curr. Sci.,* 41, 821, 1972.

75. **Phillips, D. J., MacKey, B., Ellis, W. R., and Hansen, T. N.,** Occurrence and interactions of *Aspergillus flavus* with other fungi on almonds, *Phytopathology,* 69, 829, 1979.

76. **Purcell, S. L., Phillips, D. J., and MacKey, B. E.,** Distribution of *Aspergillus flavus* and other fungi in several almond-growing areas of California, *Phytopathology,* 70, 926, 1980.

77. **Rao, A. S.,** Fungal populations in the rhizosphere of peanut *(Archis hypogaea* L.), *Plant Soil,* 17, 260, 1962.

78. **Rodriguez, J. G., Patterson, C. G., Potts, M. F., Poneleit, C. G., and Beine, R. L.,** Role of selected arthropods in the contamination of corn by *Aspergillus flavus* as measured by aflatoxin production, in *Aflatoxin and Aspergillus flavus in Corn,* Diener, U. L., Asquith, R. L., and Dickens, J. W., Eds., Southern Cooperative Series Bulletin 279 for Southern Regional Research Project S-132, Auburn, Ala. 1983, 23.

79. **Rodriguez, J. G., Potts, M., and Rodriguez, L. D.,** Survival and reproduction of two species of stored product beetles on selected fungi, *J. Invertebr. Pathol.,* 33, 115, 1979.

80. **Russell, T. E., Von Bretzel, P., and Easley, J.,** Harvesting method effects on aflatoxin levels in Arizona cottonseed, *Phytopathology,* 71, 359, 1981.

81. **Russell, T. E., Lee, L. S., and Bucl, S.,** Seasonal formation of aflatoxins in cottonseed produced in Arizona and California, *Plant Dis.,* 71, 174, 1987.

82. **Russell, T. E., Watson, T. F., and Ryan, G. F.,** Field accumulation of aflatoxin in cottonseed as influenced by irrigation termination dates and pink bollworm infestation, *Appl. Environ. Microbiol.,* 31, 711, 1976.

83. **Sanders, T. H., Blankenship, P. D., Cole, R. J., and Hill, R. A.,** Effect of soil temperature and drought on peanut pod and stem temperatures relative to *Aspergillus flavus* invasion and aflatoxin contamination, *Mycopathologia,* 86, 51, 1984.

84. **Sanders, T. H., Hill, R. A., Cole, R. J., and Blankenship, P. D.,** Effect of drought on occurrence of *Aspergillus flavus* in maturing peanuts, *J. Am. Oil Chem. Soc.,* 58, 966A, 1981.

85. **Sannasi, A. and Amirthavalli, S.,** Infection of the velvet mite, *Trombidium gigas* by *Aspergillus flavus, J. Invertebr. Pathol.,* 16, 54, 1970.

86. **Schroeder, H. W. and Boller, R. A.,** Invasion of rough rice in storage by strains of *Aspergillus flavus, Phytopathology,* 61, 910, 1971.

87. **Shotwell, O. L,** Aflatoxin in corn, *J. Am. Oil. Chem. Soc.,* 54, 216A, 1977.

88. **Stack, J. P. and Pettit, R. E.,** Fungi affecting the germination of sclerotia of *Aspergillus flavus* in soil, *Proc. Am. Peanut Res. Educ. Soc.,* 17, 71, 1985.

89. **Stack, J. P. and Pettit, R. E.,** Colonization of organic matter substrates in soil by *Aspergillus flavus, Proc. Am. Peanut Res. Educ. Soc.,* 16, 45, 1984.

90. **Stephenson, L. W. and Russell, T. E.,** The association of *Aspergillus flavus* with hemipterous and other insects infesting cotton bracts and foliage, *Phytopathology,* 64, 1502, 1974.

91. **Tomas, M. J. E., Pujol, M. D. S., Congregado, F., and Suarez-Fernandez, G.,** Method to assess antagonism of soil microorganisms towards fungal spore germination, *Soil Biol. Biochem.,* 12, 197, 1980.

92. **Tsao, C. H. and Berisford, Y. C.,** Infection in the bagworm larvae by *Aspergillus parasiticus, Bull. Ga. Acad. Sci.,* 34, 52, 1976.

93. **Van Walbeek, W., Clademenos, T., and Thatcher, F. S.,** Influence of refrigeration on aflatoxin production by strains of *Aspergillus flavus, Can. J. Microbiol.,* 15, 629, 1969.

94. **Weckbach, L. S. and Marth, E. H.,** Aflatoxin production by *Aspergillus parasiticus* in a competitive environment, *Mycopathologia,* 62, 39, 1977.

95. **Wehner, F. C. and Rabie, C. J.,** The micro-organisms in nuts and dried fruits, *Phytophylactica,* 2, 165, 1970.

96. **Wells, J. M. and Payne, J. A.,** Toxigenic species of *Penicillium, Fusarium,* and *Aspergillus* from weevil-damaged pecans, *Can. J. Microbiol.,* 22, 281, 1976.

97. **Wells, J. M. and Payne, J. A.,** Toxigenic *Aspergillus* and *Penicillium* isolates from weevil-damaged chestnuts, *Appl. Microbiol.,* 30, 536, 1975.

98. **Wicklow, D. T.,** Taxonomic features and ecological significance of sclerotia, in *Aflatoxin and Aspergillus flavus in Corn,* Diener, U. L., Asquith, R. L., and Dickens, J. W., Eds., Southern Cooperative Series Bulletin 279 for Southern Regional Research Project S-132, Auburn, Ala., 1983, 6.

99. **Wicklow, D. T. and Hesseltine, C. W.,** Fluorescence produced by *Aspergillus flavus* in association with other fungi in autoclaved corn kernels, *Phytopathology,* 69, 589, 1979.

100. **Wicklow, D. T., Hesseltine, C. W., Shotwell, O. L., and Adams, G. L.,** Interference competition and aflatoxin levels in corn, *Phytopathology,* 70, 761, 1980.

101. **Wicklow, D. T. and Shotwell, O. L.,** Intrafungal distributions of aflatoxins among conidia and sclerotia of *Aspergillus flavus* and *Aspergillus parasiticus, Can. J. Microbiol.,* 29, 1, 1983.

102. **Wicklow, D. T. and Wilson, D. M.,** Germination of *Aspergillus flavus* sclerotia in a Georgia maize field, *Trans. Br. Mycol. Soc.,* 87, 651, 1986.

103. **Widstrom, N. W., McMillian, W. W., and Wilson, D. M.,** Contamination of preharvest corn by aflatoxin, Proc. 39th Annu. Corn and Sorghum Research Conf., American Seed Trade Association, Washington, D.C., 1985, 68.

104. **Wilson, D. M. and Bell, D. K.,** Aflatoxin production by *Aspergillus flavus* and *A. parasiticus* on visibly sound dehydrated peanut, corn and soybean seed, *Peanut Sci.,* 11, 43, 1984.

105. **Wilson, D. M. and Flowers, R. A.,** Unavoidable low level aflatoxin contamination of peanuts, *J. Am. Oil Chem. Soc.,* 55, 111A, 1978.

106. **Wilson, D. M., Gueldner, R. C., and Hill, R. A.,** Effects of plant metabolites on the *Aspergillus flavus* group and aflatoxin production, in *Aflatoxin and Aspergillus flavus in Corn,* Diener, U. L., Asquith, R. L., and Dickens, J. W., Eds., Southern Cooperative Series Bulletin 279 for Southern Regional Research Project S-132, Auburn, Ala., 1983, 13.

107. **Wilson, D. M., McMillian, W. W., and Widstrom, N. W.,** Use of *Aspergillus flavus* and *A. parasiticus* color mutants to study aflatoxin contamination of corn, in *Biodeterioration VI,* Llewellan, G., Rear, C. O., and Barry, S., Eds., International Biodeterioration Society, London, 1986, 284.

108. **Wilson, D. M., McMillian, W. W., and Widstrom, N. W.,** Differential effects of *Aspergillus flavus* and *A. parasiticus* on survival of *Heliothis zea* (Boddie) and *Spodoptera frugiperda* (J. E. Smith) (Lepidoptera: Noctuidae) reared on inoculated diet, *Environ. Entomol.,* 13, 100, 1984.

109. **Wilson, D. M. and Stansell, J. R.,** Effect of irrigation regimes on aflatoxin contamination of peanut pods, *Peanut Sci.,* 10, 54, 1983.

Chapter 6

CONTROL OF TOXIGENIC MOLDS IN CEREAL SEEDS

Raul G. Cuero, Alison Murray, and John E. Smith

TABLE OF CONTENTS

I. INTRODUCTION

As early as 1876 Tyndall[65] referred to "the struggle for existence" among mixed cultures of bacteria and fungi in organic media. That antagonism could exist between microorganisms was well demonstrated by Pasteur and Joubert[52] who successfully repressed the production of anthrax in susceptible animals by simultaneous inoculation with *Bacillus anthracis* and other bacteria.

In natural substrates such as roots, leaves, cereal seeds, etc., microorganisms are present in variable numbers and are in a constant state of interaction with individual microorganisms competing for nutrients and space. Such interactions will be influenced by the nature of the substrate, water activity, temperature, gaseous changes, etc., and will promote an ecological succession of microorganisms. In most cases, the total numbers of organisms or quantity of biomass will be limited by the protective qualities of the living system, e.g., seed testa and leaf epidermis. However, any damage to these protective surfaces will allow microbial passage to the rich nutrient environment within the structure, with subsequent qualitative and quantitative changes in the microbial flora. In many cases this can lead to extensive or total decay by saprophytic, facultative, or parasitic microorganisms.

The economy of the world depends heavily on the production and utilization of cereals not only for human and animal food and feed requirements, but also for numerous chemical and biotechnological conversion processes. Table 1 gives some indication of the various production levels of the most commonly grown cereals and how these have changed over the last two decades.[40] Grain production will be influenced by the vagaries of climate and soil together with the ever constant predation of insects, nematodes, and microorganisms. In the present context only one group of microorganisms, the filamentous fungi, will be considered, and further within this group only those filamentous fungi which produce mycotoxins will be considered.

Whereas most other microbial attacks on cereals can often lead to serious detrimental physico-chemical changes in the substrates, this may not always be the case with mycotoxin producing fungi. Grossly contaminated cereals, where mixed microbial cultures are present, rarely, if ever, contain mycotoxins. Mycotoxin production in nature is most often associated with the near monoculture existence of the toxigenic fungus; and again, levels of contamination need not be high to have mycotoxins present.

II. MYCOTOXINS AND MYCOTOXICOSES

Mycotoxins are, in general, low molecular weight, nonantigenic fungal metabolites capable of eliciting a toxic response in man and animals. Mycotoxins are synthesized typically, but not exclusively, in cereal grains, nuts, and other plant materials. Mycotoxins are typical fungal secondary metabolites and are formed by a consecutive series of enzyme-catalyzed reactions from a few biochemically simple intermediates of primary metabolism, e.g., acetate, mevalonate, malonate, and certain amino acids.[61,64]

The principal pathways involved in the formation of mycotoxins are, viz., the polyketide route (e.g., aflatoxins), the amino route (e.g., roquefortine), the terpene route (e.g., trichothecenes), and the tricarboxylic acid route (e.g., rubratoxin).

Mycotoxicosis is poisoning by the ingestion of mycotoxins in food altered or damaged by the growth of toxigenic molds. Acquisition by the host is usually by ingestion, though inhalation or direct skin contact can represent health hazards.[67,26] At high concentrations many mycotoxins can produce acute disease syndromes while, at lower levels, they can be carcinogenic, mutagenic, teratogenic, or oestrogenic, they may reduce the growth rate of young animals, and they can interfere with native mechanisms of resistance and impair immunologic responsiveness making animals more susceptible to other microbial infections.[54]

Mycotoxins can be produced by fungi growing in the living plant, decaying plant material,

Table 1
TREND IN WORLD GRAIN PRODUCTION 1961 TO
1979

Cereal crop	Mean of 1975-79 ($\times 10^9$ tons)	Change during last two decades (%)
Cereals total	1499	+ 34
Wheat	408	+ 38
Rice paddy	366	+ 31
Maize	352	+ 39
Barley	176	+ 43
Sorghum	62	+ 42
Oats	49	+ 2
Millet	40	+ 5
Rye	33	− 1
Mixed cereals and others	12	+ 27

From MacKey, B., *A Renewable Resource,* Pomeranz, Y. and Munck, L., Eds., American Society of Cereal Chemists, St. Paul, Minn., 1981, 5. With permission.

and in stored plant material.[28] In the present context the growth of toxigenic fungi in maize and rice grains will be examined; in particular, the interactive roles of substrates, temperature, water activity, and time will be considered together with the impact of cocultivation with toxigenic and nontoxigenic fungi. Such studies may give some insight into the reasons for the well documented fluctuations in mycotoxin occurrence in growing and stored cereals.

III. GRAIN MYCOFLORA

Although the external surfaces of cereal grains, before and after harvest, will contain a wide and numerous range of microorganisms, only a small number will be able to invade the grain kernel. Several groups of fungi can invade the kernels while they are developing on the plant or after they have matured. In wheat and barley the principal invasive field fungi are *Alternaria, Fusarium, Helminthosporium,* and *Cladosporium. Alternaria* is present in nearly 100% of wheat and barley kernels (generally not visible to the naked eye), and will normally do no further damage unless the harvested grains are kept unduly moist. Increasingly it is being recorded that *Aspergillus flavus* invades maize kernels in the field often after there has been insect damage.

The fungi that successfully colonize cereal grains in the field invariably require high moisture levels in the kernels; grain maturation coupled with efficient drying treatments will reduce moisture levels and a_w (water activity) resulting in the inhibition or death of these fungi.[13] However, many of these field fungi are potent mycotoxin producers, in particular the *Fusarium* spp., and although the producer fungi will eventually die, the mycotoxins will remain mostly undamaged by most drying conditions.

During harvest and subsequent transportation and storage, other mold species such as *Aspergillus* and *Penicillium* will settle onto the grain surfaces and may be able to germinate and grow at the now reduced a_w levels. Many of the well-documented toxigenic fungi, and their toxins, can cause serious contamination of stored cereals, e.g., *A. flavus* and *A. parasiticus* (aflatoxins), *P. viridicatum* (ochratoxin, citrinin, penicillic acid), and many others.

Although filamentous fungi are the most well documented of microorganisms colonizing cereal grains, many bacteria and yeasts have also been identified and play a dominant role in advanced biodeterioration. Recent unpublished studies by the authors have shown that the presence of specific yeasts or bacteria can strongly influence aflatoxin production in maize seeds exposed to specific environmental conditions. It is probable that the presence of the bacteria influence the nutritive condition of the maize stimulating aflatoxin production. To what extent this may occur in nature is not yet well understood.

IV. INTRINSIC AND EXTRINSIC FACTORS REGULATING MICROBIAL GROWTH AND MYCOTOXIN FORMATION

Specific qualities, inherent in the substrate, that can influence microbial growth and metabolic activity include chemical, physical, and biochemical characteristics (nutrient composition, natural antimicrobial factors, pH, a_w, Eh, etc.). Subtle interactions of these factors will determine the selection and successful growth of those microorganisms present on or in the cereal grains. Organisms unable to compete in a particular environment will gradually be eliminated, leaving the substrate to more dominant or tolerant species.

The physical nature of the seed coat has been shown to have a restrictive effect on mold penetration, colonization, and mycotoxin production.[12] Maize varieties have been shown to vary in their ability to support *A. flavus* growth and toxin production, most probably due to variable thickness and strength of pericarps.[37,71]

Intraseed colonization by toxigenic fungi can be strongly influenced by the chemical nature of the substrate. Cereals with high carbohydrate and lipid content tend to favor aflatoxin production. The effect of different heat sterilized substrates on aflatoxin production was early realized by Hesseltine et al.[27] who showed that strains of *A. flavus* varied widely in their ability to produce aflatoxin in rice, sorghum, peanuts, maize, wheat, and soybeans. Soybeans were a particularly poor substrate for aflatoxin synthesis. Eugenio et al.[21] showed that zearalenone production by *Fusarium roseum* was higher on polished rice than on maize and wheat, with even less on oats and barley and none on soybeans and peas. However, Richardson et al.[55] have recently demonstrated zearalenone production in soybeans and enhanced production of the trichothecene HT-2 in soybean meal compared to rice culture.

Accumulation of metabolic by-products or waste products could limit both the producer organism and accompanying organisms. In contrast, such products may serve as substrates for specific organisms allowing them to achieve higher stability in their environment. Such interactions may be synergistic or antagonistic, according to their growth enhancing or inhibitory nature.

It has been considered that the mycotoxins released by toxigenic fungi may function *in situ* as antimicrobial compounds thus permitting a selective advantage to the toxigenic fungus. This, however, seems unlikely since most mycotoxins, such as the aflatoxins, ochratoxin A, and zearalenone, have little in vitro antimicrobial activity. Such mycotoxins that do have observable antimicrobial activity, e.g., citrinin, have not been extensively demonstrated in nature in quantities likely to provide an obvious competitive advantage to the producer fungus.[10,38,69]

Secondary metabolism, as exemplified by micotoxins, is not well understood, but is generally linked with the growth of the organism and is most probably a part of the normal maturation process of that microorganism.[9] Secondary metabolism, or the secondary metabolites, have been limited with morphological and physiological differentiation.[4,7] Secondary metabolites are formed almost invariably as a response of the microorganism to specific growth conditions, which experimentally have been characterized by some form of nutrient limitation. Such observations have been made generally in defined media under specific controlled environmental conditions. In contrast, in the rich milieu of a cereal seed, nutrient limitations seems somewhat unlikely. Table 2 lists some of the suggested concepts for secondary metabolism.[32] Ciegler[14] has suggested that mycotoxins may also act as chemical signalling agents for higher organisms.

Table 2
CONCEPTS FOR SECONDARY METABOLISM

- Shunt products are synthesized during the idiophase.
- Breakage of regulatory mechanisms and changes in cell permeability are necessary.
- Overflow metabolism is a consequence of nutrient limitation.
- Secondary metabolism is growth linked. Energy-consuming secondary metabolism will decrease with increasing maintenance coefficients.
- Secondary metabolism is connected with differentiation.
- Secondary metabolites act as regulator molecules and secondary metabolism increases flexibility of cells to adapt to changing environments.
- Specific molecules regulate secondary metabolism.
- Secondary metabolism is a playground of evolution.

From Hutter, R., *Bioactive Microbial Products: Search and Discovery,* Bu'Lock, J. D., Nisbet, L. J., and Winstanley, D. J., Eds., Academic Press, London, 1983, 37. With permission.

In the natural environment, mycotoxins generally occur in a sporadic nature and there is little or no evidence to indicate how often the producer organism grows in a natural substrate without producing toxins. Similarly, it is not known if a specific toxin is produced at all times, but is metabolized by co-existing microorganisms.

V. INTERRELATIONSHIP BETWEEN WATER ACTIVITY AND TEMPERATURE ON MYCOTOXIN FORMATION

It is a general observation that temperature and water activity strongly interact for mycotoxin production, particularly in conditions of grain storage. Storage conditions have a profound effect on fungal growth and mycotoxin formation. *Aspergillus flavus* is one of the most common storage fungi. Significant invasion of grains by storage fungi occurs only after harvest.[29,30]

An $a_w < 0.60$, equivalent to a moisture content of below 13% in starchy cereal seeds such as wheat, barley, rice, maize, and sorghum, below 12% in soybeans, and below 10% in flax-seeds, prevents invasion by storage fungi regardless of how long the grains are stored. As the a_w rises above these levels, invasion by storage fungi increases. Storage temperatures between 4.0 and 10°C will greatly decrease the rate of growth of storage fungi, but will not prevent it completely.

Growth, sporulation, and concomitant toxin production can be regulated by controlling temperature. At high temperatures *A. flavus* can grow at relatively low a_w, and mycelial growth occurs between 6 and 46°C optimum. However, maximum growth is not necessarily conducive to high aflatoxin production. Marth[44] reported that aflatoxin is produced only between 13 and 41°C, with an optimum temperature of 28°C.

Aspergillus parasiticus has an optimum temperature of 35°C for growth, but only forms aflatoxin in maximum concentrations between 25 and 30°C in 7 to 21 days both on semisynthetic and natural substrates, e.g., moist sorghum, groundnuts, and cotton seeds.[51]

Lin et al.[39] demonstrated that cycling of temperatures between 33 and 15°C favored aflatoxin B_1 accumulation, whereas cycling between 25 and 15°C favored aflatoxin G_1 formation. Cultures subject to temperature cycling between 33 and 25°C at various time intervals changed the relative production of aflatoxin B_1 and G_1. They suggested that the enzyme system responsible for the conversion of aflatoxin B_1 to G_1 might be more efficient at 25°C than at 35°C.[62] Borroughs and Sauer[8] reported a marked succession of fungi, including toxigenic species, in high-moisture grain, particularly at 25°C. For example, the percentage of seeds yielding *Fusarium* increased at 20 to 24% moisture content; *Penicillium* grew at 18 to 24% moisture content; *Trichothecium* grew at 22 to 24% moisture content; but *Alternaria* decreased with increasing moisture content and time. They also found that under such conditions of high moisture content and at 25°C, a toxin producing fungus would grow and produce toxins and then die. If the grains were tested for known toxin-producing fungi, the fungus might be absent or nearly so, while its toxin might still be present.

Sauer and Borroughs,[56] working with mixtures of wet and dry maize, found that *A. flavus* grew rapidly and produced aflatoxin in blends (mixtures) of maize with mean moisture contents (MC) of about 17.5% and a_w of 0.86 to 0.87, while blends with a_w below 0.85 (18.5% MC) generally had limited *A. flavus* growth and no aflatoxin production. Blends with average water content below 17% (0.82 a_w) were invaded primarily by the *A. glaucus* group. *Penicillium* invaded most of the maize above 16% moisture content (0.80 a_w).

In another series, Diener and Davis,[19,20] using peanuts, found that the minimum conditions for aflatoxin production by *A. flavus* was 0.85 a_w for 21 days at 30°C. The limiting low temperature for visible growth and aflatoxin production by the fungus was 13°C for 21 days at 0.97 to 0.99 a_w. The limiting relative humidity at 20°C for sound and broken mature kernels was 83% (0.83 a_w), whereas it was 86% (0.86 a_w) for immature kernels and 92% (0.92 a_w) for kernels from unshelled peanuts.

Similarly, Chang and Markakis[11] found high aflatoxin production by *A. parasiticus* in barley containing >16% water, and also more aflatoxin in hulled barley than entire seeds.

Although the growth of *A. flavus* is greatly affected by temperature, aflatoxin production is not necessarily affected in the same way. Schindler et al.[57] found that aflatoxin production was not related to the growth rate of *A. flavus*; one isolate at 41°C, at almost maximal growth of *A. flavus*, produced no aflatoxins. He also found that the ratio of the production of aflatoxin B_1 to aflatoxin G_1 varied with temperature.

The effects of water content and temperature on the production of other mycotoxins has been reported. Yields of zearalenone on maize were increased when *Fusarium* spp. were incubated at 24 to 27°C for 7 days or 11 to 14°C for 4 to 6 weeks. Sherwood and Peberdy[59] also reported maximum amounts of zearalenone from *F. graminearum* at 10 to 14°C. Although a lower temperature is usually preferred for toxin production by most Fusaria, *F. graminearum* produced more zearalenone in sorghum grain at 25°C than at 10°C, and optimum conditions on maize were at least 21°C and 9 days incubation.[22]

Eugenio et al.,[21] working with maize and rice, found more zearalenone was produced by *F. graminearum* at a moisture content of 60 to 65% in rice, and in maize at a moisture of 45% w/w basis.

Climatic conditions may influence the growth and toxicity of *Fusarium* strains. Marasas et al.[43] found *F. moniliforme* to be most prevalent in northern South Africa which is warmer, drier, and more tropical. Also, Cuero et al.[16] reported a clear influence of climate on the ecological distribution of toxigenic fungi in the highlands and lowlands of Colombia. The *Fusarium* spp. and *Penicillium citrinum* were more prevalent in the highlands, while *A. flavus* and *A. ochraceus* were more prevalent in the lowlands, although all these toxigenic fungi were isolated from both agrisystems. Furthermore, the *Fusarium* spp. isolated from the highlands, with lower temperature and greater humidity, produced more zearalenone than isolates from the lowlands. Conversely, *Aspergillus* isolates from the lowlands produced more aflatoxin and ochratoxin A.

Ayerst,[3] working with 30 isolates from 12 species of *Aspergillus*, three species of *Penicillium,* and *Stachybotrys atra* found that all isolates were most tolerant of low water activity at temperatures close to the optimum, and maximum temperatures were higher at reduced water activity.

Northolt and Bullerman[49] found that a natural substrate, not completely equilibrated to the intended a_w, can give experimental error. Most investigations allow equilibration of the substrate above liquids of known a_w, with a_w produced from the salt and its concentration, and without measurement of the actual a_w. Making allowance for such factors, they found that the minimum a_w for growth and mycotoxin production was 0.07 to 0.10 a_w lower than found by other workers.

Niles[48] studied growth and aflatoxin production of *A. flavus* in wheat and barley. Both growth and aflatoxin production were restricted by low temperature, low a_w, and high temperature, but not by high a_w. At 15°C, the lowest temperature at which growth occurred, 0.95 a_w was needed for growth, while at 42°C growth was severely restricted and aflatoxin production altogether inhibited. The optimum temperature for growth was 35°C, at lower temperatures (15°C, 20°C) the optimum a_w was 0.975, but at temperatures above these, it was 0.95.

Water activity-temperature relationships of the growth of fungi, other than *A. flavus*, and their respective toxin production has also been studied. Spore germination of *Alternaria alternata*, *Cladosporium herbarum*, *Verticillium lecanii*, the *Aspergillus* spp., *A. amstelodami*, *A. fumigatus*, the *Penicillium* spp., *P. ficeum*, and *P. roqueforti*, occurred at a lower a_w than linear growth which was again at a lower a_w than asexual sporulation. In other experiments, Magan and Lacey[41] also showed that a Numerical Index of Dominance (I_D) between interacting field and storage fungi varied with water activity, temperature, and substrate. Thus, variation in temperature between 15 and 30°C was accompanied by changes in I_D scores. Substrate also affected I_D, and total I_D scores of the most dominant fungi were usually lower on wheat extract than on the richer malt agar.

Production of the toxins altenuene (AE), alternariol (AOH), and alternariol monomethyl ether (AME) by *Alternaria alternata* on wheat was affected by both water activity and temperature. They reported greater production of all three mycotoxins at 0.98 a_w and 25°C. Little toxin was produced at 0.90 a_w.[42]

Northolt et al.[50] demonstrated the profound effect of water activity and temperature relationship on growth and patulin production by *Penicillium* and *Aspergillus* strains. They found that some species, *A. ochraceus* and *P. expansum*, only formed the toxins at high a_w (0.99) under laboratory conditions, whereas in apples under similar conditions of temperature, the production of patulin varied, although they did not specify the a_w of the apples. They suggested a high a_w requirement for patulin production.

VI. MICROBIAL INTERACTIONS

Mycotoxin production has been studied mostly in monoculture on rich, autoclaved, organic media under clearly defined and controlled environmental conditions. There have been few attempts to study the growth of toxigenic fungi and mycotoxin formation in mixed microbial culture, in either synthetic or natural substrates, or in fluctuating enviromental conditions.

Development of *Aspergillus flavus* and aflatoxin is regulated by competition with other fungi that invade peanut kernels, especially *A. niger* and *Rhizoctonia solani*.[2] However, prior inoculation by the *A. niger* was necessary to inhibit *A. flavus*.

Similarly, simultaneous inoculation of autoclaved rice with *A. chevalieri* or *A. candidus* decreased the quantity of aflatoxins produced by *A. parasiticus*, while significant quantities of aflatoxins were detected when *A. parasiticus* was inoculated alone.[5,6] They suggested that the decrease in aflatoxin production was caused by inhibition of *A. flavus* by the dominant *A. candidus* and *A. chevalieri*, rather than by the metabolism of the toxin. Conversley, Alderman et al.[1] reported an increase of aflatoxin in a mixed culture of *A. flavus* with *P. italicum* or *Lactobacillus planetarum*.

The effect of fungal interaction on aflatoxin formation has also been studied in autoclaved maize. Inoculating with pairs of cultures of either *A. niger* or *Trichoderma viride* with *A. flavus*, Wicklow and Hesseltine[70] found no aflatoxin production when *A. flavus* was paired with either *A. niger* or *T. viride*. However, aflatoxin production was detected when *A. flavus* was paired with *Candida guillermondii*. From the effects of dual inoculation of aflatoxin and kojic acid production, they also argued that the competitor species determined which biosynthetic pathways were available to *A. flavus*, since kojic acid could be produced without aflatoxins in the solid substrate fermentations.

A few experiments have utilized growing plants. *A. flavus* colonization of immature and mature pericarps on gnotobiotically grown peanut plants was increased by *Trichoderma viride*, but *P. funiculosum* nullified this effect. On almond kernels or hulls, *Ulocladium chartarum* was the most effective antagonist of *A. flavus*, although most of the fungi inoculated onto almond hulls before harvest decreased colonization of the kernels by *A. flavus* or *A. parasiticus* at high a_w.[53]

A very frequent association between *A. flavus* and *F. moniliforme* in freshly harvested maize, without effect on aflatoxin production, has been reported.[16,17,23]

One of the most widely recognized competitive interactions in cereal grains and other substrates is that between *A. flavus* Link ex Fr. and *A. niger*.[2,36,58,60,63,68] *A. niger* can inhibit either the growth of *A. flavus* or aflatoxin production on maize, peanut, or rice seeds. Aflatoxin was not degraded by *A. niger* but its production was inhibited by the decreased pH resulting from *A. niger* colonization.[31]

Moss and Badii[46,47] studied the effect of rubratoxin B, a metabolite produced by *P. rubrum* and *P. purpurogenum*, on growth, metabolism, and morphology of *A. niger* and *A. parasiticus*. They showed abnormal mycelium formation in *A. niger* and repression of *A. parasiticus* growth, but increased aflatoxin production. Alderman et al[1] previously reported that rubratoxin B inhibited growth of *A. parasiticus* and enhanced aflatoxin production.

The potential for microbial growth and toxin formation will be a combination of intrinsic and extrinsic environmental factors coupled with the genetic properties of the microorganisms; thus, in a particular environment, an organism grows in a characteristic manner and at a characteristic rate. While microorganisms remain metabolically active they will continue to interact. Dominance within a microflora will not be permanent but rather dynamic. Interactions can be synergistic or antagonistic depending on whether the effect is growth enhancing or inhibitory.

VII. MYCOTOXIN PRODUCTION IN ASEPTIC LIVING CEREAL SEEDS

In almost all of the previously discussed examples of mycotoxin production in cereal substrates, the substrate had been autoclaved and contained under aseptic experimental conditions in flasks or similar containment systems. In the authors' laboratory an experimental protocol has been developed which ensures microorganism-free cereal seeds. A containment system that readily permits environmental control of water activity, temperature, and aeration while retaining aseptic operation was also developed. This system has now been used successfully to study the interactions between toxigenic fungi and other microorganisms, in particular filamentous fungi, yeasts, and bacteria.[15,18]

A. Microbial-free Seeds

Surface sterilization of seeds, etc. has regularly been carried out by means of washing in sodium hypochlorite solutions. Such treatments can efficiently remove surface contaminants but do not affect internally sited microorganisms.[24] For most experimental studies, cereal seeds would normally be autoclaved for variable periods of time and then retained under aseptic conditions. However, this treatment drastically changes the chemical and physical nature of the substrate such that it no longer fully resembles the previous living seeds. Therefore, extrapolation of experimental results obtained in these autoclaved systems can bear no similarity to natural living systems.

An alternative means of achieving total kill of contaminating microorganisms is by treatment with gamma irradiation. The susceptibility of microorganisms to radiation has been extensively studied[33,34] and a wide range of susceptibilities has been demonstrated.[35,45]

In the present series of experiments maize seeds were equilibrated to 22% moisture content at 4°C for 2 weeks prior to exposure to various irradiation dosages.[18] A specific radiation

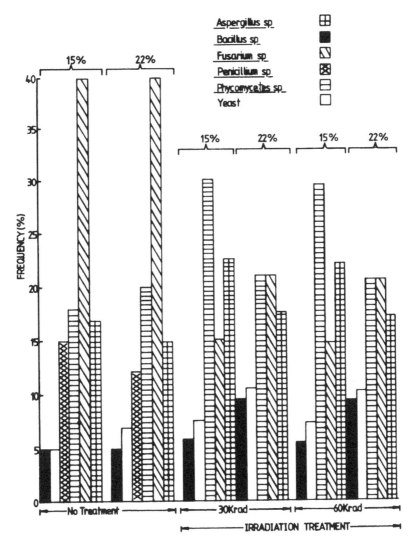

FIGURE 1.   Effect of irradiation treatment on the frequency of isolation of the *Aspergillus* sp., *Bacillus* sp., *Fusarium* sp., *Pencillium* sp., *Phycomycetes* sp., and yeast colony types from maize of different water contents. (From Cuero, R., Smith, J., and Lacey, J., *J. Food Microbiol.*, 3, 2, 1986. With permission.)

dosage of 1200 krad eliminated all seed-borne microorganisms while still retaining the ability of the seed to germinate. Microorganisms varied in their sensitivity to gamma irradiation (Figures 1, 2, and 3). The aspergilli and the penicilli were eliminated by 120 krad. The Phycomycetes were eliminated at 600 krad, while yeast and the *Fusarium* spp. were particularly resistant and were only eliminated by treatment at 1200 krad. Low dosages of irradiation also appeared to stimulate germination of the maize seeds.

B. The Containment System

The experimental system utilized Fison Environmental Cabinets® with temperature and humidity controls in which the containment bags with the irradiated seeds and inoculum were placed.[15] The bags were constructed from Valmic microporous film, a specially formulated opaque white polypropylene material with micropores less than 0.4 μm. This microporous material was initially developed for mushroom spawn production by the French spawnmaker Somycel and the material has been submitted for patenting.

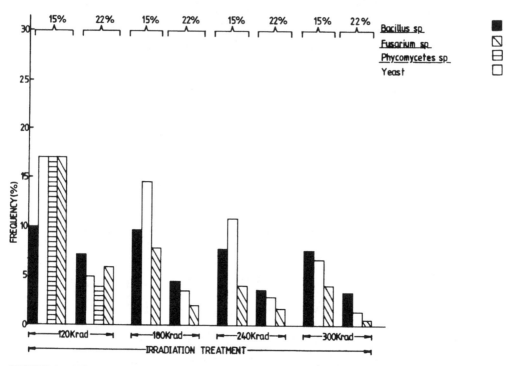

FIGURE 2.   Effect of irradation treatment on the frequency of isolation of the *Bacillus* sp., *Fusarium* sp., *Phycomycetes* sp., and yeast colony types from maize of different water contents. (From Cuero, R., Smith, J., and Lacey, J., *J. Food Microbiol.*, 3, 2, 1986. With permission.)

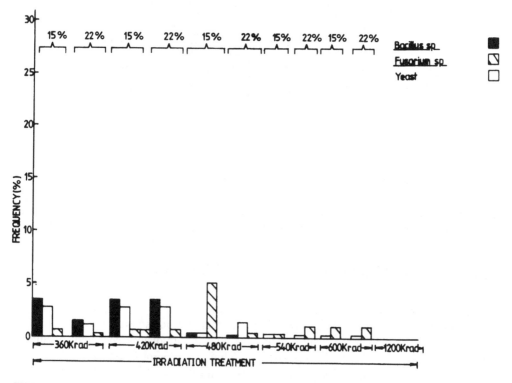

FIGURE 3.   Effect of irradation treatment on the frequency of isolation of the *Bacillus* sp., *Fusarium* sp., and yeast colony types from maize of different water contents. (From Cuero, R., Smith, J., and Lacey, J., *J. Food Microbiol.*, 3, 2, 1986. With permission.)

Table 3
GROWTH OF ORGANISMS ON CRACKED MAIZE AT DIFFERENT TEMPERATURES AND A_w AFTER 12 DAYS

	Temperature (25°C)			Temperature (16°C)		
	0.98 a_w	0.95 a_w	0.90 a_w	0.98 a_w	0.95 a_w	0.90 a_w
A. flavus	4	4	3	4	—	NDG
A. oryzae	4	3	2	4	—	NG
A. niger	4	3	2	3	—	NG
P. viridicatum	4	4	3	4	1	—
F. graminearum	4	3	2	2	—	NG

Note: Macroscopic observation of apparent mold growth using a semiquantitative scale was carried out: (—) very little growth; (1) $1/4$ of the substrate covered; (2) $1/2$ of the substrate covered; (3) $3/4$ of the substrate covered; (4) the whole substrate completely covered; (NG) no growth; (NDG) no distinctive growth.

The microporous bags can be reused with repeated autoclaving. In the present system, the size of the bags was 13 × 23 cm, and they were sealed by impulse heating. The most important feature of the bag is the Valmic microporous film which allows efficient air (ca. 25 cm³/cm²/min at 1 atm) and moisture vapor transmission (ca. 500 g/m²/24 hr at 23°C and 50% RH) with no cross movement of fungal spores. Throughout the experiments (1 to 2 weeks) internal RH (relative humidity) and temperature values as determined by an electronic dew point meter (Protometer DP 680) were relatively constant, reaching steady values not significantly different from the external conditions in less than 2 hr.

C. Mycotoxin Production in Monoculture and Mixed Culture Systems Under Various Water Activity/Temperature Conditions

In the following series of experiments, maize[72], previously sterilized by gamma irradiation at 1200 krad, was contained aseptically in the microporous bags in the Fision Cabinets® set at RH levels to give a_w values of 0.98, 0.95, and 0.90 and at two temperatures, viz. 25 and 16°C.

Three toxigenic fungi were used in these studies, viz., *Aspergillus flavus*, *Penicillium viridicatum*, and *Fusarium graminearum*, producing, respectively, aflatoxin, ochratoxin A, and zearalenone. In separate studies each toxigenic fungus was paired with each of the toxigenic fungi or with other nontoxigenic fungi, viz., *Aspergillus niger* and *Aspergillus oryzae*. Growth of the single cultures and the mixed cultures was monitored visually and toxin analyses were carried out at the completion of each experiment, normally 2 weeks.

1. Interaction of Fungi

Table 3 shows the growth patterns of all individual fungi on the cracked maize at the various a_w values and temperatures. Mold growth was assessed using a semiquantitative scale.

All fungi were able to grow at 25°C over all water activity levels. In contrast, at 16°C most showed reduced growth or even no observable growth over 0.95 to 0.90 a_w. All organisms grew well at 0.98 a_w.

In cocultivation *A. flavus* showed either dominance (marked visible growth) or mutual inhibition (moderate visible growth) with the other fungi over the whole range of water activity levels and both temperatures (Table 4). However, *P. viridicatum* showed striking dominance over *A. flavus* at 16°C at all a_w values.

Table 4
INTERACTIONS BETWEEN *A. FLAVUS* AND OTHER FUNGI ON CRACKED MAIZE AT DIFFERENT TEMPERATURES AND A_w AFTER 12 DAYS

Temperature a_w	Temperature (25°C)			Temperature (16°C)		
	0.98	0.95	0.90	0.98	0.95	0.90
A. oryzae	4(AOR)	3(AOR)	2(AOR)	1(AOR)	2(AOR)	NG(AOR)
	4(AFL)	3(AFL)	2(AFL)	1(AFL)	2(AFL)	NG(AFL)
A. niger	3(ANG)	3(ANG)	2(ANG)	1(ANG)	2(ANG)	1(ANG)
	4(AFL)	4(AFL)	3(AFL)	1(AFL)	3(AFL)	NG(AFL)
P. viridicatum	1(PVR)	1(PVR)	−(PVR)	4(PVR)	4(PVR)	2(PVR)
	4(AFL)	3(AFL)	3(AFL)	−(AFL)	−(AFL)	NG(AFL)
F. graminearum	4(AFL)	4(AFL)	3(AFL)	1(AFL)	3(AFL)	NG(AFL)

Note: Macroscopic observation of apparent mold growth using a semiquantitative scale was carried out: (−) very little growth; (1) $^1/_4$ of substrate covered; (2) $^1/_2$ of substrate covered; (3) $^3/_4$ of substrate covered; (4) the whole substrate completely covered. (NG) no growth; (NDG) no distinctive growth; (AFL) *A. flavus;* (AOR) *A. oryzae;* (ANG) *A. niger;* and (PVR) *P. viridicatum.*

2. Mycotoxin Production in Monocultures

Inoculated maize was analyzed for the various mycotoxins after 12 days incubation at the various a_w levels and two temperatures. Aflatoxin production increased with increasing a_w in maize (Table 5). Production was highest at 0.98 a_w at 25°C, while none was formed at 16°C and 0.90 a_w. *P. viridicatum* failed to produce ochratoxin A on maize in monocultures over the whole a_w/temperature combination. With *F. graminearum*, zearalenone production increased with increasing a_w at 25°C while production at 16°C was maximum at 0.95 a_w.

3. Aflatoxin Production in Paired Culture of A. flavus and the other Toxigenic and Nontoxigenic Fungi

The results of these pairing experiments are shown in Figures 4 and 5.

a. Aspergillus flavus Paired with A. oryzae

Aflatoxin production at 0.98 a_w and 25°C was approximately unchanged when compared to monocultures. However, at 0.95 a_w there was a considerable decrease in aflatoxin production while at 0.90 a_w there was a measurable increase in aflatoxin production.

These two molds have a close taxonomic relationship and may compete for similar nutrients and for the same ecological niche. The reduction of the aflatoxin level could have been by a mutualistic inhibition on growth rate or by interference. It is probable that *A. oryzae* competes more effectively at the higher a_w value than *A. flavus*. Griffin[25] reported that the more effective antagonists of *A. flavus* growing on almonds were those fungi requiring high to moderate moisture levels.

b. Aspergillus flavus Paired with A. niger

A. niger only decreased aflatoxin formation by *A. flavus* at 0.98 a_w, 25°C. By contrast, it increased aflatoxin production at 0.90 to 0.95 a_w at 25°C. Similarly, aflatoxin was greater at 0.95 to 0.98 a_w and 16°C than in a single culture under the same conditions. It is possible that because *A. niger* can invade the substrate faster at both optimum and minimum conditions for fungal growth it had rendered the substrate more favorable for *A. flavus* growth and aflatoxin production, perhaps by substrate breakdown or by modifying pH. At 0.98 a_w and 25°C *A. niger* may have either restricted growth of *A. flavus* affecting its morphogenesis,

Table 5
TOXIN PRODUCTION BY SINGLE
TOXIGENIC CULTURES IN MAIZE
KERNELS EXPOSED TO
DIFFERENT WATER ACTIVITY
AND TEMPERATURE CONDITIONS

Organism and toxin	a_w		
	0.98	**0.95**	**0.90**
Aspergillus flavus (aflatoxins)			
25°C	1020.0	989.0	100.2
16°C	56.0	73.6	0.0
Penicillium viridicatum (ochratoxin A)			
25°C	N.D.	N.D.	N.D.
16°C	N.D.	N.D.	N.D.
Fusarium graminearum (zearalenone)			
25°C	404.0	310.0	138.0
16°C	289.0	453.0	0.0

Note: Values given are ng/g. (N.D.) toxin not detected.

e.g., sporulation or germination, or its metabolism. Perhaps the ratio between these two *Aspergillus* species was altered thus affecting aflatoxin formation as indicated by Hill et al.[30] They found that when the ratio of *A. flavus* to *A. niger* exceeded 19:1, aflatoxin was present, but when it was less than 19:1, no aflatoxin could be detected. Most previous workers have found decreased aflatoxin production by *A. flavus* in mixed culture with *A. niger* in autoclaved substrate at high water content and at 25°C.[31,60] Most of these studies in autoclaved substrates do not report exact values of a_w; however, in this laboratory, autoclaved grain reached an a_w around 0.98. So, these results on aflatoxin production by *A. flavus* paired with *A. niger* in autoclaved grains agree with the present results. Nevertheless, previous workers have not reported the increase of aflatoxin by *A. flavus* paired with *A. niger* in grain. The gnotobiotic seeds and the wider range of conditions used in this study resembled natural conditions more closely.

The regular pattern of aflatoxin production over the whole range of environmental conditions may be a consequence of the similar growth pattern of *A. flavus*.

Therefore, the effect of *A. niger* on aflatoxin production by *A. flavus* in maize kernels is likely to be more related to physicochemical changes in the substrate caused by the faster growth of *A. niger* at 0.90 to 0.95 a_w and 25°C. While at 0.98 a_w and 25°C, *A. niger* may have caused morphological changes to *A. flavus* which resulted in the reduction of *A. flavus* cells with further reduction of aflatoxin production.

c. Aspergillus flavus Paired with P. viridicatum

P. viridicatum enhanced aflatoxin production by *A. flavus*. Possibly *P. viridicatum* increased aflatoxin production by enhancing growth of *A. flavus*. Wells et al.[66] found that *P. funiculosum* not only nullified the antagonistic effect of *Trichoderma viride* on *A. flavus* but also appeared to have a direct stimulating effect on colonization of gnotobiotically grown peanuts by *A. flavus*. The increasing aflatoxin production towards the lowest a_w values at 25°C suggests that it is unlikely to be an a_w-temperature effect, although the effect may be altered by a_w and temperature. Thus, it is also likely that at lower a_w and 25°C *P. viridicatum*

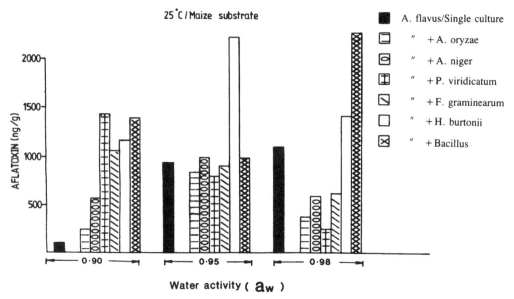

FIGURE 4.   Aflatoxin production by *A. flavus* in single or in dual culture with other colonists in cracked maize kernels after 12 days of incubation at 25°C and 0.90, 0.95, and 0.98 a$_w$.

has introduced physicochemical changes to the substrate, rendering it more favorable than *A. flavus* for growth and consequent aflatoxin production; perhaps *P. viridicatum* has also released elicitors or precursors of aflatoxin formation. High aflatoxin production in peanuts at a lower water content has been reported.[29]

The decreased aflatoxin production at the highest a$_w$ at 25°C in maize could be a metabolic effect of *P. viridicatum* causing perhaps an antibiotic effect, or it could be as a result of enzyme action by *P. viridicatum* acting as an antagonist.

Therefore, from the results of this study, *P. viridicatum* affects the formation of aflatoxin, possibly by inhibiting mechanisms of aflatoxin formation rather than the growth of *A. flavus*. Further studies using a substrate and conditions in which both fungi are able to produce their respective toxins will be needed in order to confirm the antibiotic effect, and perhaps to consider the enzymatic action.

d. Aspergillus flavus Paired with Fusarium graminearum

The pattern of aflatoxin production by *A. flavus* interacting with *F. graminearum* at different a$_w$s and temperatures, relatively, was similar to that obtained with *P. viridicatum* except that more aflatoxin was produced at 25°C than with *P. viridicatum*. Hence a similar discussion as for pairing with *P. viridicatum* can be given for *A. flavus* paired with *F. graminearum*. *F. graminearum* possibly enhances aflatoxin production in the same way as *P. viridicatum*. It is likely that the competitor *F. graminearum*, at 25°C, releases metabolites or alters the substrate, making some nutrients or precursor substances more readily available for *A. flavus* growth and/or for aflatoxin formation. Furthermore, it is possible that *F. graminearum*, at 25°C in maize, stimulates morphological changes of *A. flavus*, e.g., increased mycelial growth, which could lead to an increase of aflatoxin production by *A. flavus*. *A. flavus* and *F. moniliforme* were frequently associated in uninoculated maize kernels and *F. moniliforme* always grew faster than *A. flavus*.[16,17] However, no antagonism with *A. flavus* nor any effect on aflatoxin production was found.

The noticeable increase of aflatoxin production by *A. flavus* when paired with *F. graminearum* at 0.95 to 0.90 a$_w$ at 25°C, as compared to aflatoxin production by monoculture

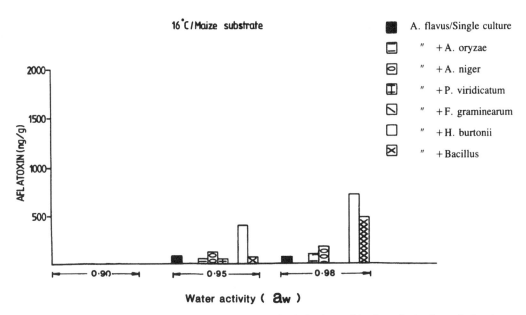

FIGURE 5. Aflatoxin production by *A. flavus* in single or in dual culture with other colonists in cracked maize kernels after 12 days of incubation at 16°C and 0.90, 0.95, and 0.98 a_w.

of *A. flavus*, could be a result of the increased growth of *A. flavus* stimulated by the greater metabolic activity of *F. graminearum* under the conditions studied (in maize). At 16°C and 0.90 a_w this association did not stimulate aflatoxin production or fungal growth. Therefore, it is likely that temperature affects the effectiveness of the association between these two fungi to produce aflatoxin by *A. flavus*. The regular pattern of aflatoxin production over the whole range of environmental conditions may be a consequence of the similar growth pattern of *A. flavus*.

These results warrant further studies on the effect of metabolites such as alcohols, lipids, acids, etc. produced by *F. graminearum* on aflatoxin production by *A. flavus*.

VIII. CONCLUDING REMARKS

A clear and comprehensive understanding of the fundamental base line of microbial ecology is necessary in order to have a practical application of mixtures of organisms as biological control, especially for metabolites such as mycotoxins. The practical implications of mixed culture must be assessed in vitro before being put into practice although it is always difficult to try to simulate natural habitats in the laboratory. However, the advantage of laboratory approach is that it is possible to control the appropriate parameters and examine the effect of a single parameter on the population under study.

ACKNOWLEDGMENTS

It is with deep sorrow that we report that Miss Alison Murray, a young and highly talented microbiologist, was tragically killed during the preparation of this paper.

REFERENCES

1. **Alderman, G. G., Emeh, C. O., and Marth, E. H.,** Aflatoxin and rubratoxin produced by *Aspergillus parasiticus* and *Penicillium rubrum* when grown independently, associatively or with *P. italicum* or *Lactobacillus planetarum*, *Z. Lebensm. Unters. Forsch.,* 153, 305, 1973.
2. **Ashworth, L. J., Schroeder, H. W., and Langley, B. C.,** Aflatoxins: environmental factors governing occurrence in Spanish peanuts, *Science,* 14, 1228, 1965.
3. **Ayerst, G.,** The effects of moisture and temperature on growth and spore germination in some fungi, *J. Stored Prod. Res.,* 5, 127, 1969.
4. **Berry, D. R.,** The environmental control of the physiology of filamentous fungi, in *The Filamentous Fungi,* Vol. 1, Smith, J. E. and Berry, D. R., Eds., Edward Arnold, London, 1975, 16.
5. **Boller, R. A. and Schroeder, H. W.,** Influence of *Aspergillus chevalieri* on production of aflatoxin in rice by *Aspergillus parasiticus, Phytopathology,* 63, 1507, 1973.
6. **Boller, R. A. and Schroeder, H. W.,** Influence of *Aspergillus candidus* on production of aflatoxin in rice by *Aspergillus parasiticus, Phytopathology,* 64, 121, 1974.
7. **Bu'Lock, J. R.,** Secondary metabolism in fungi: its relationship to growth and development, in *The Filamentous Fungi,* Vol. 1, Smith, J. E. and Berry, D. R., Eds., Edward Arnold, London, 1975, 33.
8. **Burroughs, R. and Sauer, D. B.,** Growth of fungi in sorghum grain stored at high moisture contents, *Phytopathology,* 61, 767, 1971.
9. **Calam, C. T.,** Secondary metabolism as an expression of microbial growth and development, *Folia Microbiol.,* 24, 276, 1979.
10. **Campbell, I. M.,** Secondary metabolism and microbial physiology, *Adv. Microb. Physiol.,* 25, 2, 1984.
11. **Chang, H. G. and Markakis, P.,** Effect of moisture content on aflatoxin production in barley, *Cereal Chem.,* 58, 89, 1981.
12. **Chelkowski, J., Dopierala, G., Godlewska, B., Radomyska, W., and Szebiotko, K.,** Mycotoxins in cereal grain. Production of ochratoxin in different varieties of wheat, rye and barley, *Nahrung,* 25, 625, 1981.
13. **Christensen, C. M. and Kauffmann, H. H.,** Microflora, in *Storage of Cereal Grains and Their Products,* Christensen, C. M., Ed., American Association of Cereal Chemists, St. Paul, Minn., 1974, 158.
14. **Ciegler, A.,** Do mycotoxins function in ecological processes?, *J. Food Safety,* 5, 23, 1983.
15. **Cuero, R. G., Smith, J. E., and Lacey, J.,** A novel containment system for laboratory scale solid particulate fermentations, *Biotech. Lett.,* 7, 463, 1985.
16. **Cuero, R. G., Hermandez, I., Cardenes, H., and Osorio, E.,** Produccion de micotoxinas en maiz de distintas zones del Valle del Cauca, Colombia, in *Aflatoxin in Maize: Proceedings of the Workshop,* Zuber, M. S., Lillehoj, E., and Renfro, B. L., Eds., International Maize and Wheat Improvement Center, Mexico, 1986.
17. **Cuero, R. G., Lillehoj, E. B., Kwolek, W. F., and Zuber, M. S.,** Mycoflora and aflatoxin in preharvest maize kernels of varied endosperm type, in *Trichothecenes and Other Mycotoxins,* Lacey, J., Ed., John Wiley & Sons, New York, 1986, 109.
18. **Cuero, R. G., Smith, J. E., and Lacey, J.,** Influence of gamma irradiation and sodium hypochlorite on microflora and germination of maize seeds, *J. Food Microbiol.,* 3, 2, 1986.
19. **Diener, U. L. and Davis, N. D.,** Limiting temperatures and relative humidity for growth and production of aflatoxin and free fatty acids by *Aspergillus flavus* in sterile peanuts, *J. Am. Oil Chem. Soc.,* 44, 259, 1967.
20. **Diener, U. L. and Davis, N. D.,** Limiting temperatures and relative humidity for aflatoxin production by *Aspergillus flavus* in stored peanuts, *J. Am. Oil Chem. Soc.,* 47, 347, 1970.
21. **Eugenio, C. P., Christensen, C. M., and Mirocha, C. J.,** Factors affecting production of the mycotoxin F-2 by *Fusarium roseum, Phytopathology,* 60, 1055, 1970.
22. Food and Agriculture Organization, *Global Perspective on Mycotoxins,* Food and Nutrition paper, MYC-13-4a, 1979.
23. **Fennel, D. I., Bothast, R. J., Lillehoj, E. B., and Peterson, R. E.,** Bright greenish-yellow fluorescence and associated fungi in white corn naturally contaminated with aflatoxin, *Cereal Chem.,* 50, 404, 1973.
24. **Flannigan, B.,** Enumeration of fungi and assay for ability to degrade structural and storage components of grain, in *Biodeterioration Investigation Techniques,* Walters, A. H., Ed., Applied Science, London, 1977.
25. **Griffin, D. M.,** Soil moisture and ecology of soil fungi, *Biol. Rev.,* 38, 141, 1963.
26. **Hayes, A. W.,** Mycotoxins: biological effects and their role in human diseases, *Clin. Toxicol.,* 17, 45, 1980.
27. **Hesseltine, C. W., Shotwell, O. L., Ellis, J. J., and Stubblefield, R. D.,** Aflatoxin formation by *Aspergillus flavus, Can. J. Microbiol.,* 18, 631, 1966.

28. **Hesseltine, C. W.,** Conditions leading to mycotoxin contamination of foods and feeds, in *Mycotoxins and Other Fungal Related Food Problems,* Rodericks, J. V., Ed., Advances in Chemistry Series No. 149, American Chemical Society, Washington, D.C., 1976, 146.

29. **Hill, R. A., Blankenship, P. D., Cole, R. J., and Sanders, T.,** Effect of soil moisture and temperature on preharvest invasion of peanuts by the *Aspergillus flavus* group and subsequent aflatoxin development, *Appl. Environ. Microbiol.,* 45, p. 1986, 1983.

30. **Hill, R. A., Wilson, D. M., McMillan, W. W., Cole, R. J., Sanders, T. H., and Blankenship, P. D.,** Ecology of the *Aspergillus flavus* group and aflatoxin formation in maize and groundnuts, in *Trichothecenes and Other Mycotoxins,* Lacey, J., Ed., John Wiley & Sons, New York, 1986, 79.

31. **Horn, B. W. and Wicklow, D. T.,** Factors influencing the inhibition of aflatoxin production in corn by *Aspergillus niger, Can. J. Microbiol.,* 29, 1087, 1983.

32. **Hutter, R.,** Design of culture media capable of provoking wide gene expression, in *Bioactive Microbial Products: Search and Discovery,* Bu'Lock, J. D., Nisbet, L. J., and Winstanley, D. J., Eds., Academic Press, London, 1983, 37.

33. International Atomic Energy Agency — Food and Agriculture Organization, Training manual on food irradiation technology and techniques, Technical Reports Series No. 114, International Atomic Energy Agency, Vienna, 1970.

34. International Atomic Energy Agency — Food and Agricultural Organization, *Food Preservation by Irradiation,* Vol. 1, Proceedings of a Symposium, International Atomic Energy, Vienna, 1978.

35. **Ito, H., Iizuka, H., and Sato, T.,** Identification of osmophilic *Aspergillus* isolated from rice and their radiosensitivity, *Agric. Biol. Chem.,* 37, 789, 1973.

36. **Joffe, A. Z.,** Relationships between *Aspergillus flavus, A. niger* and other fungi in the mycoflora of groundnut kernels, *Plant and Soil,* 31, 57, 1969.

37. **King, S. B.,** Time of infection of maize kernels by *Fusarium moniliforme* and *Cephalosporium acremonium, Phytopathology,* 71, 796, 1980.

38. **Lillehoj, E. B.,** Secondary metabolites as chemical signals between species in an ecological niche, in *Advances in Biotechnology, Vol. III Fermentation Products,* Vezina, C. and Lingh, K., Eds., Pergamon Press, Elmsford, N. Y., 1980, 397.

39. **Lin, Y. C., Ayres, J. C., and Koehler, P. E.,** Influence of temperature cycling on the production of aflatoxins B_1 and G_1 by *Aspergillus parasiticus, Appl. Environ. Microbiol.,* 40, 333, 1980.

40. **MacKey, B.,** Cereal products, in *A Renewable Resource: Theory and Practice,* Pomeranz, Y. and Munck, L., Eds., American Association of Cereal Chemists, St. Paul, Minn., 1981, 5.

41. **Magan, N. and Lacey, J.,** Effect of water activity, temperature and substrate on interaction between field and storage fungi, *Trans. Br. Mycol. Soc.,* 82, 83, 1984.

42. **Magan, N., Gayley, G., and Lacey, J.,** Effect of water activity and temperature on mycotoxin production by *Alternaria alternata* in culture and on wheat grain, *Appl. Environ. Microbiol.,* 47, 1113, 1984.

43. **Marasas, W. F. O., Kriek, N. P. J., Wiggins, W. M., Steyn, P. S., Towers, D. K., and Hastie, T. J.,** Incidence, geographic distribution and toxigenicity of *Fusarium* spp. in South African corn, *Phytopathology,* 69, 1181, 1979.

44. **Marth, E. H.,** Mycotoxins — A review, 7th Annu. Symp. N.Y. Agricultural Experimental Station, Special Report 13, 1973, 13.

45. **Mohyuddin, M. and Skoropad, P.,** Effect of ^{60}Co gamma irradiation on the survival of some samples of each of three different grades of wheat, *Can. J. Bot.,* 48, 217, 1970.

46. **Moss, M. O. and Bandii, F.,** Effect of rubratoxin A on growth, metabolism and morphology of *Aspergillus niger, Trans. Brit. Mycol. Soc.,* 74, 1, 1980.

47. **Moss, M. O. and Bandii, F.,** Increased production of aflatoxin by *A. parasiticus* Speare, in the presence of rubratoxin B, *Appl. Environ. Microbiol.,* 43, 895, 1982.

48. **Niles, E. V.,** Growth and aflatoxin production of *Aspergillus flavus* in wheat sterilised by gamma irradiation, ethylene oxide fumigation and autoclaving, *Trans. Br. Mycol. Soc.,* 70, 239, 1978.

49. **Northolt, M. D. and Bullerman, L. L.,** Prevention of mould growth and toxin production through control of environmental conditions, *J. Food Prot.,* 65, 519, 1982.

50. **Northolt, M. D., van Egmund, H. P. and Paulsch, W. E.,** Patulin production by some fungal species in relation to water activity and temperature, *J. Food Prot.,* 41, 885, 1978.

51. **Park, K. H. and Bullerman, L. B.,** Increased aflatoxin production by *Aspergillus parasiticus* under conditions of cycling temperature, *J. Food Sci.,* 46, 1147, 1981.

52. **Pasteur, L. and Joubert, J.,** Charbon et septiceme, *C. R. Acad. Sci.,* 85, 101, 1877.

53. **Phillips, D. J., MacKey, B., Ellis, W. R., and Hansen, T. N.,** Occurrence and interaction of *Aspergillus flavus* with other fungi on almonds, *Phytopathology,* 69, 829, 1979.

54. **Pier, A. C., Richard, J. L., and Cysewski, S. J.,** Implications of mycotoxins in animal disease, *J. Am. Vet. Med. Assoc.,* 176, 719, 1980.

55. **Richardson, K. E., Hagler, W. M., Hanney, C., and Hamilton, P. B.,** Zearalenone and trichothecene production in soybeans by toxigenic *Fusarium* spp., *J. Food Prot.,* 48, 240, 1984.

56. **Sauer, D. B. and Burroughs, R.,** Fungal growth, aflatoxin production and moisture equilibrium in mixtures of wet and dry corn, *Phytopathology,* 70, 516, 1980.
57. **Schindler, F., Palmer, J., and Eisenberg, W. V.,** Aflatoxin production by *Aspergillus flavus* as related to various temperatures, *Appl. Microbiol.,* 15, 1006, 1967.
58. **Semenuik, G.,** Storage of cereal grains and their products, in *Microflora,* Anderson, J. A. and Alcock, A. W., Eds., American Association of Cereal Chemists, St. Paul, Minn., 1954.
59. **Sherwood, R. F. and Peberdy, J. F.,** Production of the mycotoxin zearalenone by *Fusarium graminearum* growing on stored grain. I. Grain storage at reduced temperatures, *J. Sci. Food Agric.,* 25, 1081, 1974.
60. **Shotwell, O. L., Goulden, M. L., Bothast, R. J., and Hesseltine, C. W.,** Mycotoxins in hot spots in cereal grains. I. Aflatoxin and zearalenone occurrence in stored grain, *Cereal Chem.,* 52, 687, 1975.
61. **Steyn, P. S., Ed.,** *The Biosynthesis of Mycotoxins: A Study in Secondary Metabolism,* Academic Press, London, 1980, 432.
62. **Stutz, H. K. and Krumperman, P. J.,** Effect of temperature cycling on the production of aflatoxin by *Aspergillus parasiticus, Appl. Environ. Microbiol.,* 32, 327, 1976.
63. **Tsubsuehi, H., Yamamoto, K., Hisada, K., Sakade, Y., and Tsuchihiva, K.,** Degradation of aflatoxin B_1 by *Aspergillus niger, Proc. Jpn. Assoc. Mycotoxicol.,* 12, 33, 1981.
64. **Turner, W. B. and Aldridge, D. C.,** *Fungal Metabolites,* Academic Press, London, 1983, 11.
65. **Tyndall, J.,** The optical deportment of the atmosphere in relation to the phenomenon of putrefaction and infection, *Philos. Trans. R. Soc. London,* 166, 27, 1876.
66. **Wells, T. R., Kreutzer, W. A., and Lindsey, D. L.,** Colonisation of gnotobiotically grown peanuts by *Aspergillus flavus* and selected interacting fungi, *Phytopathology,* 62, 1238, 1972.
67. **World Health Organization,** *Environmental Health Critiera. II. Mycotoxins,* World Health Organization, Geneva, 1979.
68. **Wicklow, D. T.,** Adaptation in wild and domesticated yellow-green Apergilli, in *Toxigenic Fungi — Their Toxins and Health Hazard,* Kurata, M. and Ueno, Y., Eds., Elsevier, N.Y., 1983.
69. **Wicklow, D. T. and Carrol, G. C.,** *The Fungal Community: Its Organisation and Role in the Ecosystem,* Marcel Dekker, New York, 1981.
70. **Wicklow, D. T. and Hesseltine, C. W.,** Fluorescence produced by *Aspergillus flavus* in association with other fungi in autoclaved corn kernels, *Phytopathology,* 69, 589, 1979.
71. **Zuber, M. S., Calvert, O. H., Lillehoj, E. B., and Kivolek, F.,** Preharvest development of aflatoxin B_1 in corn in USA, *Phytopathology,* 66, 1120, 1976.
72. **Cuero, R. G., Smith, J. E., and Lacey,** Stimulation by *Hyphopichia burtonii* and *Bacillus amyloliquefasciens* of aflatoxin production by *Aspergillus flavus* in irradiated maize and rice grains, *Appl. Environ. Microbiol.,* 53(5), 1142, 1987.

Chapter 7

AFLATOXIN CONTAMINATION IN MAIZE AND ITS BIOCONTROL

M. S. Zuber and E. B. Lillehoj

TABLE OF CONTENTS

I. INTRODUCTION

Prior to 1960, most of the research on aflatoxin contamination in agricultural commodities was confined to stored grain, where at that time it was considered a most important problem. In 1960, turkey deaths in England were found to be caused by aflatoxin-contaminated feed made from moldy ground-nut meal from Brazil. Aflatoxin-contaminated foodstuffs were subsequently associated with a high incidence of liver cancer in humans, especially in the tropical and subtropical countries of the world.[8,13,56] Since 1960 there has been an explosion in the amount of aflatoxin research. A major thrust of the work has been an effort to reduce or eliminate aflatoxin contamination by acquiring more information on the biology of the toxin-producing species, *Aspergillus flavus* and *A. parasiticus*, as well as on factors affecting fungal infection and the aflatoxin contamination level.

II. PREHARVEST OCCURRENCE

Rambo et al.[70] reported the occurrence of *A. flavus* in preharvest maize in 1971 and 1972. They found that some types of weather conditions were associated with elevated preharvest infections. Several authors[4,47,73,96] have reported preharvest occurrences of *A. flavus* from Iowa to the southern U.S. In 1977, low levels of aflatoxin were found in isolated lots of maize grain from Illinois, Indiana, Iowa, and Missouri.[52] In the same year, in the southeastern U.S., some lots of freshly-harvested maize grain were found to have 2000 ppb aflatoxin, and many contained over 200 ppb.[73]

The potential for widespread occurrence of aflatoxin in preharvest maize grain became a reality in 1977, when significant amounts of aflatoxin contamination were found in a large geographic area of the U.S. Samples from harvested grain across the southern Corn Belt states of Alabama,[30] Georgia,[60,95] and North Carolina[38] showed aflatoxin levels exceeding 20 ppb in more than half of the samples from each state. A systematic sampling of four hybrids over four planting dates at one site in each of nine states showed very high levels of contamination in Florida, Georgia, and South Carolina; high levels in Missouri; moderate amounts in Tennessee, Kansas, and Iowa; and low levels in Ohio and Illinois.[52] McMillian et al.[61] reported results from a 6-year study (1977 to 1982) showing that aflatoxin contamination in Georgia was a problem every year, but varied greatly in degree from year to year (Table 1). Aflatoxin contamination was heavy in 1977 and 1980, moderate in 1981, and light in 1978, 1979, and 1982.

III. FACTORS AFFECTING INFECTION AND AFLATOXIN LEVELS

Maize genotypes grown at different locations across the central and southern Corn Belt demonstrated that the incidence and level of aflatoxin were consistently much greater in the southern U.S. than in the Corn Belt.[52] Weather patterns and ear-damaging insects were the two major differences between northern and southern growing conditions, so researchers began to study how these two factors related to aflatoxin contamination.

A. Climate

The dry hot growing seasons of 1977, 1980, and 1981 in the southeastern U.S. were also seasons of high incidence of aflatoxin contamination. *A. flavus* has been reported[20,36] to be a thermotolerant fungus, thus making it more competitive at high temperatures. In laboratory studies, Trenk and Hartman[79] found that temperatures in the 35°C range did not restrict aflatoxin formation. Lillehoj et al.[49,51] compared summer temperature to the number of aflatoxin positives for 12 genotypes grown at 9 locations in 1976, and 12 locations in 1978 (Table 2). In general, the locations with the highest July to August temperature had the

Table 1
AFLATOXIN CONTAMINATION AND VISIBLE *A. FLAVUS* SPORULATION RECORDED IN PREHARVEST DENT CORN IN THE GEORGIA COASTAL PLAIN FROM 1977 THROUGH 1982

Year	Ears with visible *A. flavus*[a] (%)	Aflatoxin- contaminated samples (%)	Range in sample levels of aflatoxin[b] (ng/g)	Mean level of aflatoxin[a] (ng/g)	Insect damaged ears[a,c] (%)
1977	18 A	100	0—4708	78 A	100 A
1978	2 BC	85	0—620	8 B	60 C
1979	1 C	74	0—3029	6 B	55 C
1980	5 B	100	2—1616	77 A	56 C
1981	2 BC	84	0—634	30 A	99 A
1982	1 C	57	0—1730	5 B	79 B

[a] Any two means in a column not followed by the same letters are significantly different from one another at the 1% level.
[b] The sum of $B_1 + B_2 + G_1 + G_2$ aflatoxins.
[c] Geometric means (the antilogarithm of the logarithmic means) of the sum of $B_1 + B_2 + G_1 + G_2$ aflatoxins.

From McMillian, W. W., Wilson, D. M., and Widstrom, N. W., *J. Environ. Qual.*, 14, 200, 1985. With permission.

Table 2
SUMMER TEMPERATURES AND A TYPICAL DISTRIBUTION OF AFLATOXIN-POSITIVE SAMPLES FOUND AMONG TWELVE COLLECTED FROM EACH OF NINE LOCATIONS IN 1976 AND AMONG TWELVE ENTRIES OF TWO REPLICATIONS COLLECTED FROM TWELVE LOCATIONS IN 1978[49,51]

Location	1976		1978	
	No. aflatoxin- positive samples	Av. July—Aug. temp. (°F)	No. aflatoxin- positive entries	Av. aflatoxin concentration (ng/g)
Florida	9	81	12	50
Georgia	2	78	12	83
Texas	1	81	12	18
North Carolina	1	77	12	266
Missouri	1	77	12	53
Mississippi	1	76	11	11
Iowa	0	73	4	5
Illinois	0	73	0	0
Ohio	0	67	1	Trace
Kansas	—	—	11	63
South Carolina	—	—	12	224
Tennessee	—	—	12	44

highest number of aflatoxin-positive samples in preharvest corn. In 1978, locations with the highest concentration of aflatoxin were North Carolina, South Carolina, and Georgia; whereas, the more northern Corn Belt states of Iowa and Illinois had a very low incidence of contamination. Gray et al.[30] reported increased amounts of aflatoxin within regions of Alabama in 1977, corresponding with lower than normal rainfall that year. Jones et al.[41] showed that

Table 3
SIMPLE CORRELATIONS
BETWEEN AFLATOXIN
CONTAMINATION OF GRAIN
SAMPLES AND WEATHER
TRAITS RECORDED ON THE
GEORGIA COASTAL PLAIN
IN MAY THROUGH AUGUST
FROM 1977 TO 1982

Weather trait	Correlation with aflatoxin concentration (ng/g)
Mean temperature (°C)	0.85[a]
Relative humidity (%)	−0.62
Total net evaporation (cm)	0.97[b]
Mean radiation (Langleys/day)	−0.57
Total precipitation (cm)	0.24

[a]   Significantly different from zero at the 5% level of probability.
[b]   Significantly different from zero at the 1% level of probability.

From McMillian, W. W., Wilson, D. M., and Widstrom, N. W., *J. Environ. Qual.*, 7, 564, 1978. With permission.

the process of ear infection favored temperatures above 32°C. McMillian et al.[61] found, during a 6-year study, that both mean temperature and total net evaporation were significantly and positively correlated with levels of field contamination (Table 3). Thompson et al.[77] studied the effects of day/night regimes, thermal units per day, and three stages of kernel development for inoculating maize kernels from plants grown in controlled-environment chambers. Highest aflatoxin levels were found for the 30-day/26-night temperature regimes, when inoculated at 21 days post-flowering (Table 4). These results clearly showed that high temperatures during the grain filling period increased aflatoxin levels.

B. Role of Insects

The increased occurrence of *A. flavus* contamination when insects invade the maize ear is well documented. Taubenhaus[76] was the first to report a specific association between *A. flavus* infection and insect damage, also mentioning that insect injury to maize increased in dry weather. The significant ear-damaging insects include corn earworm, *Heliothis zea*, European corn borer, *Ostrinia nubilalis*, rice weevil, *Sitophilus oryzae*, and maize weevil, *Sitophilus zeamais*. This association was further established by Widstrom,[91] and McMillian et al.[60] How important a role insects play has been associated with the region, the growing season (whether favorable or stressful), and the insect species involved. Ear-damaging insects can carry *A. flavus* spores into the ear externally and internally. In vectoring studies, Widstrom et al.[89] showed that corn earworm damage gave the highest number of aflatoxin-positive ears but that damage from the European corn borer yielded three times more aflatoxin

Table 4
MEAN CONCENTRATION (PPB) OF AFLATOXIN B_1 IN TWO
CORN CULTIVARS INOCULATED WITH *ASPERGILLUS FLAVUS*
AT THREE KERNEL-DEVELOPMENT STAGES AND GROWN IN
FOUR POSTINOCULATION TEMPERATURE REGIMES

		Postinoculation temp. regime day/night (°C)				
		22/18	30/18	22/26	30/26	
Kernel development stage at inoculation	Cumulative no. of thermal units (°C)	Thermal units/day (°C)[b]				Mean
		9.5	12.5	14.5	17.5	
Gaspe						
Early-dough	133.0	7,834	3,890	5,649	6,918	5,875
Medium-dough	255.5	8,241	12,705	13,121	16,293	12,218
Late hard-dough	378.0	13,646	28,040	35,075	48,641	28,642
Mean		9,572	11,272	13,740	17,620	12,753
LSR:[a]	Individual					2.30
	Kernel stage					1.52
	Postinoculation					1.62
W103 × Gaspe						
Early-dough	245.0	12,647	12,445	12,972	12,705	12,677
Medium-dough	367.5	11,194	22,029	26,485	29,444	20,941
Late hard-dough	434.0	13,092	17,989	25,823	31,046	20,849
Mean		12,303	17,022	20,701	22,646	17,701
LSR:[a]	Individual					1.74
	Kernel stage					1.32
	Postinoculation					1.37

[a]   (LSR) means whose ratios exceed the LSR values are significantly different, $P < 0.05$.
[b]   All plants grown for about 365 cumulative thermal units after inoculation.

Modified from Thompson, D. L., Lillehoj, E. B., Leonard, K. J., Kwolek, W. F., and Zuber, M. S., *Crop Sci.,* 20, 609, 1980.

B_1 (Table 5). A study by Fennell et al.[28] showed that the European corn borer was a more effective vector than the corn earworm on ears augmented with *A. flavus* spores and non-inoculated ears (Table 6).

In a search for a suitable vector to be used as an inoculation tool Barry et al.,[11] compared two arthropod vectors using *A. flavus* spores on developing maize ears. The two vectors were the wheat curl mite, *Eriophye trilipae*, and the maize weevil, *Sitophilus zeamais*. The mite did not effectively vector *A. flavus* spores into the ear (Table 7), but the maize weevil dusted with *A. flavus* spores effectively transferred the fungus into developing ears.

In general, it has been shown that ear-damaging insects can play a role in *A. flavus* infection and subsequent aflatoxin contamination. However, not all the variation in aflatoxin levels has been explained by insect damage ratings. Correlations between aflatoxin levels and insect damage gave r values of 0.53.[100]

Table 5

**PROPORTIONS OF AFLATOXIN-POSITIVE
SAMPLES AND THE TOXIN
CONCENTRATIONS ASSOCIATED WITH
INFESTATION OF GEORGIA FIELD PLOTS
BY THREE INSECT SPECIES IN 1974**

Insect	Aflatoxin-positives from 32 plots	Aflatoxin B$_1$[a] concentration
Corn earworm	17	22
European corn borer	13	65
Fall armyworm	9	32

[a] Geometric means in ng/g.

From Widstrom, N. W., Sparks, A. N., Lillehoj, E. B., and Kwolek, W. F., *J. Econ. Entomol.*, 68, 855, 1975. With permission.

Table 6

**AFLATOXIN CONCENTRATION IN 1976 MISSOURI AND
GEORGIA CORN INFESTED WITH CORN EARWORM
AND EUROPEAN CORN BORER**

Treatment	Aflatoxin B$_1$ (ng/g)[a]	
	Noninoculated	*A. flavus* inoculated
Control	11	34
European corn borer (5—10 days)[b]	9	95
European corn borer (20 days)	50	233
Corn earworm (7—10 days)	26	196

[a] LSF = 5.7 = Least statistically significant ratio between two means (5% level).
[b] Infested at given number of days postflowering.

From Fennell, D. I., Lillehoj, E. B., Kwolek, W. F., Gutherie, W. D., Sheely, R., Sparks, A. N., Widstrom, N. W., and Adams, G. L., *J. Econ. Entomol.*, 71, 624, 1978. With permission.

IV. POSTHARVEST OCCURRENCE

Development of storage fungi in a postharvest commodity is determined by a number of factors: (1) availability of inoculum, (2) physical integrity of the seed, (3) moisture, (4) temperature, (5) aeration, and (6) nature of the substrate. Species of *Aspergillus* and *Penicillium* are common storage fungi, routinely found in stored maize grain with 13 to 18% moisture.[18,19] Members of the *A. glaucus* group predominate at 13 to 15% moisture, but above 15% other fungi can occur, including the aflatoxin-producing species *A. flavus* and *A. parasiticus*. Lopez and Christensen[57] reported that *A. flavus* did not invade starch grain below 17.5% moisture. However, the distribution of moisture within a commodity is critical. Hart[34] demonstrated that the variability in moisture distribution in grain at equilibrium depends on whether the grain was absorbing or desorbing moisture during equilibration. Lillehoj et al.[48] examined the effects on susceptibility to *A. flavus* infection of blending high-moisture (27 to 28%) with low-moisture (10%) maize to mean levels of about 14%.

Table 7
INFLUENCE OF TWO ARTHROPODS AND *A.*
***FLAVUS* ON THE MEAN AFLATOXIN (B$_1$ + B$_2$)**
CONCENTRATIONS (PPB) IN KERNELS OF A
COMMON MAIZE VARIETY AT THREE
LOCATIONS

Treatment		Location		
Anthropod	*A. flavus*	Mo.	Ga.	Tenn.
None	None	6.5ab	81.5a	206.8a
None	*A. flavus*	54.9cd	208.9ab	332.5a
E. tulspae	None	3.7a	84.8a	556.7a
E. tulspae	*A. flavus*	20.8bc	54.6a	265.7a
S. zeamais	None	40.3c	151.9ab	259.3a
S. zeamais	*A. flavus*	206.4d	442.9b	416.4a
Least significant ratio		5.01	4.30	3.56

Note: Numbers in a column followed by the same letter are not significantly different ($P < 0.05$, least significant ratio).

From Barry, D., Zuber, M. S., Lillehoj, E. B., McMillian, W. W., Adams, N. J., Kwolek, W. F., and Widstrom, N. W., *Environ. Entomol.*, 14, 634, 1985. With permission.

The results demonstrated that the fungus will infect and subsequently contaminate the maize kernels despite an average moisture content below the recognized "safe" level of 15%.

Heterogeneity in stored maize may reflect preharvest crop conditions, meteorological events during harvest, or handling technique during and after harvest. Grain entering commercial channels is often intentionally blended to achieve: (1) an average moisture level required for a specific grade, (2) a specific quality level, by blending lower with higher quality grain, and (3) additional storage time, by blending dry grain with high-moisture grain. Mixing of grain provides a heterogeneous blend in which some kernels may contain enough moisture to support fungal growth. After initial development, the fungus would produce moisture of respiration, and a pocket of high moisture within the bulk of the stored lot could provide ideal conditions for a "hot spot" of prolific fungal activity.

In laboratory studies, the availability of water for microbial development is generally measured in terms of water activity (a_w); the value expresses the ratio of water vapor pressure of the substrate to the vapor pressure of pure water at equal temperature and pressure. Although stored grains are routinely examined for moisture percentage, a_w values are a superior estimate of moisture in microbial substrates because they actually present the availability of water for a developing microbe. In prepared media, water availability is controlled by the addition of appropriate quantities of salts, sugars, etc. In defined media, an optimum a_w of 0.91 to 0.99 has been observed for growth of *A. flavus* and *A. parasiticus*.[66] Although an a_w of 0.87 did not dramatically reduce fungal growth, aflatoxin production was restricted. The tests demonstrated aflatoxin production had an enhanced sensitivity to moisture levels in comparison with fungal growth processes. The studies also identified the xerotolerant rather than xerophilic properties of the toxin-producing fungi. The degree of xerotolerance in a competitive microbial environment of diminished water availability can be a key determinant in successful establishment of a specific fungus. The aflatoxin-producing species are relatively xerotolerant, but other *Aspergilli*, particularly in the *A. glaucus* group, are especially so.

Moisture requirements for growth of *A. flavus* and aflatoxin production have been examined in mature maize kernels. The fungus does not routinely exhibit extensive growth or toxin accumulation below $a_w = 0.85$.[55,71] However, at slightly higher moisture levels (0.86 to 0.87) the fungus grows rapidly and produces toxin. Laboratory studies of *A. flavus* established moisture requirements that are essential for spore germination.[9] After the spore has germinated and the mycelium has initiated development, the increased metabolic activity produces elevated levels of the moisture produced by respiratory processes. Under appropriate conditions, the increased moisture in a limited area of a stored commodity could provide ideal conditions for futher microbial development and establishment of a hot spot. Although laboratory tests have identified a relatively fastidious requirement for moisture by *A. flavus*, the inherent heterogeneity in a stored commodity can provide unique niches for microbial development in a situation that is considered "safe".[46] The ecological dynamics of a stored commodity will reflect the ability of competing microbes to colonize seed under specific sets of conditions. Competing microbial flora determine the ability of a particular species to dominate. A number of microbial species have been identified that effectively compete with aflatoxin-producing species, including *A. niger, A. oryzae*, and *R. nigricans*.[7,58,85] Biological augmentation of a competing microflora has been considered in a strategy to control aflatoxin-producing fungi in stored maize.[101]

Temperature is as important as moisture in the control of microbial development in stored maize grain. Generally, fungi grow readily between 20 and 30°C, with a broad restrictive range of 0 to 60°C.[24] The aflatoxin-producing fungi have been classified as mesophiles since they grow optimally at 36 to 38°C with a range of 6 to 46°C.[53] In an examination of the effect of temperature cycling on aflatoxin production, West et al.[87] observed a dramatic increase in toxin levels after incremental increases in temperature from 15 to 28°C. Schroeder and Hein[73] reported that exposed cultures averaging 15°C to short periods of 40 to 50°C reduced aflatoxin production. However, maintaining a 25°C average with a minimum of 10°C did not appreciably affect toxin yields. Lin et al.[55] examined a number of temperature cycling modes on aflatoxin production. Cycling between 33 and 15°C favored aflatoxin B_1 accumulation, whereas cycling between 25 and 15°C favored aflatoxin G_1. The results suggested that temperature variation was affecting the catabolic process of toxin degradation. Stutz and Krumperman[75] observed that under conditions of diurnal and nocturnal time-temperature sequencing, the total heat thermal units represented a critical aspect of the control processes. No fungal growth was observed at thermal inputs of 208 degree hours per day. Mycelial growth and production of an orange pigment were noted between 208 and 270 thermal units per day, and the fungus sporulated and produced aflatoxin above 270 degree hours per day. Temperature fluctuations in a stored commodity would clearly influence the activities of aflatoxin-producing fungi and elaboration of toxin.

For a stored commodity to support development of aflatoxin-producing fungi, a primary inoculum must be present. *A. flavus* can occur at high levels in stored maize grain.[44,69] However, other *Aspergillus* species, particularly *A. terreus*, are also commonly observed as simultaneous colonists of individual maize kernels. Other fungal species routinely observed in maize kernels include members from the genera *Penicillium, Fusarium, Rhizopus, Absidia, Mucor, Syncephalastrum*, and *Cephalosporium*.[27,44] Until the outbreaks of *A. flavus* and aflatoxin in preharvest maize were elucidated, the conventional dogma held that *A. flavus* and *A. parasiticus* were "storage fungi" because of their xerotolerance.[18,19] "Field fungi" were considered to occur in seed above 22% moisture during development, and consisted primarily of *Alternaria, Helminthosporium, Fusarium*, and *Cladosporium* spp. This distinction between "storage" and "field" fungi indicated that the relatively xerotolerant *Aspergillus* spp. would not effectively compete with the less xerotolerant "field" species, leading to the question of a large inoculum of aflatoxin-producing propagules. With the discovery of the widespread occurrence of *A. flavus* and *A. parasiticus* in preharvest maize kernels, the question regarding the source of inoculum of the species in stored grains was answered.[53]

Besides temperature and moisture conditions, other factors contribute to the vulnerability of a stored commodity to fungal attack. Aeration is a widely used procedure to dissipate "hot spots" in stored grain. Although the forced movement of air in a stored commodity can reduce the buildup of a hot spot, it also contributes to fungal development since fungi are fastidious aerobes and require oxygen for growth. The aerobic nature of the toxin-producing fungi has been used for strategies of inert-atmosphere control. Reduction of oxygen or increased levels of carbon dioxide can restrict development of toxin-producing fungi.[94]

The nature of the substrate in a stored commodity is also a determining factor in vulnerability to fungal invasion. Physical modification of seed, such as stress cracks and breakage, can provide an entry point for fungi and subsequently predispose the commodity to development of hot spots.[78] In early investigations, Koehler[42] provided evidence of a relationship between genetically determined traits in kernel pericarp thickness and susceptibility to fungal infection. Subsequent studies[64] observed genetically-mediated differences in stored maize kernels to fungal invasion. Although breeding provides a real potential for acquiring the desired characteristics to store maize grain safely, this research has not been intensively pursued. The lack of interest is probably related to the relatively cheap sources of fossil fuel that provide a way to readily dry grain. However, as fossil fuels become less available in the next few decades, new and innovative technologies for alternative types of storage will certainly encourage breeding for desired storage traits.

V. BREEDING FOR RESISTANCE

After it was established that *A. flavus* infection was occurring in preharvest maize kernels, many breeders and geneticists initiated research directed toward either eliminating or reducing aflatoxin contamination in the crop. Through host-plant resistance, breeders have had excellent success in reducing economic losses from such diseases as leaf blight, viruses, stalk, root, and ear rots, and for increasing resistance against insects such as corn earworm, European corn borer, and aphids. In almost all instances, great progress was made in breeding strategies after a source of resistance to the pest was identified. Therefore, it seems logical to employ host-plant resistance to solve the aflatoxin problem.

The first breeding effort was designed to determine if differences in aflatoxin contamination existed among hybrids grown by farmers in the U.S. The results from these studies were not conclusive,[50,90,97] although they provided information on environmental effects and other biological factors that influenced aflatoxin contamination.[44,51] Widstrom and Zuber[92] surveyed 30 studies reported in the literature that reported differences among maize genotypes for toxin levels. About two-thirds reported significant differences, but repeatability among hybrids, locations, and years was practically nonexistent. The results indicated a very high genotype by environment interaction.

The second effort involved specifically designed experiments to measure differences in aflatoxin accumulation among genotypes. There was a lack of information on inoculation methods, sampling techniques, and number of replications necessary. To obtain consistent results, kernel wounding procedures seemed necessary to elevate fungal infection levels and toxin concentrations. Initial studies used pinboards to damage kernels about 20 days post-flowering, and applied suspensions of *A. flavus* fungal spores. This provided high infection rates; however, the pinboard compromised physical barriers provided by the kernel.

Zuber et al.[98] used an 8-maize inbred line diallel to test for resistance to aflatoxin contamination. Husks on each ear were pulled back 20 days after pollination and the exposed kernels inoculated with a pinboard which damaged approximately 4 rows of 30 kernels. The injured kernels were sprayed with an *A. flavus* spore suspension. After physiological maturity, ears were harvested, dried, shelled, and assayed for aflatoxin. Results showed a highly significant general combining ability and nonsignificant specific combining ability effects. The 8 inbred lines could be grouped into low, intermediate, and high contributors of factors conditioning aflatoxin levels to their F_1 progenies. The results of this study showed that genetic factors affected levels of aflatoxin in a single-location study.

Gardner et al.[102] used a slightly modified version of the above procedure. Of the same 8 inbred lines 7 were used to make 21 possible single crosses. To determine the most effective replication number, twenty replications of single hill plots were used. Reduction in standard errors of the mean were not appreciable with more than ten replications, and six to eight replications appeared to be an optimum number. At maturity, visually infected kernels were removed from ears and assayed for aflatoxin. Genotypic differences were significant for aflatoxin, as were the variances for general and specific combining ability effects; the latter accounted for 65% of the genotype sum of squares. Rankings of aflatoxin levels for the 21 crosses that were in common with the Zuber et al. study[98] were in good agreement, with a significant r = 0.45 by the Spearman rank correlation method. The main differences between the two studies were: (1) 20 replications instead of 2, and (2) only infected kernels were analyzed in one study and aggregated samples of infected and noninfected kernels from test ears were used in the other. Widstrom et al.[93] also observed differences in aflatoxin contamination among inbreds from a nine-line dent diallele and an eight-line sweetcorn diallel.

The breeding studies showed that aflatoxin synthesis was under host genetic control but levels of toxin in genotypes with lowest levels were unacceptably high. The pinboard method of inoculation provided uniform toxin levels but concentrations far exceeded inoculation by natural contamination.

Barry et al.[12] studied the relationship between husk tightness, insect infestation, and aflatoxin contamination. Five hybrids were rated for husk tightness ranging from relatively tight to loose. Various combinations of insect infestation with European corn borer, corn earworm, and *A. flavus* spores were applied to the silks. The results demonstrated that kernels from the tight husk hybrids contained significantly less aflatoxin than did kernels from hybrids with loose husks.

Very limited study has been done on the relation of kernel composition to aflatoxin. Lillehoj[53] studied aflatoxin levels and occurrence associated with hybrids varying in endosperm characteristics at 8 locations for 2 years. Certain trends among hybrids were apparent, but results were not statistically significant. Hybrids with the highest amylose had the lowest aflatoxin content, and elevated toxin levels were observed in a waxy type with zero amylose. High lysine hybrids had higher aflatoxin levels than their nomal counterparts. Comparison of the flint and dent versions of three hybrids did not show an association with kernel type.

Since stress, such as drought and high temperatures, has been associated with increased levels of aflatoxin, the use of inherited stress-tolerant hybrids provided a host-plant potential for resistance to toxin contamination. Zuber and Lillehoj[99] suggested planting only adapted hybrids for specific regions that can usually cope with stress better than nonadapted varieties. The traits of stress tolerance and adaptation are inherited, and maize breeders have had excellent success in incorporating the pertinent characters into commercial hybrids. A greater effort should be made to develop stress endurance genotypes for the areas that have a high incidence of aflatoxin contamination.

VI. MICROBIAL ECOLOGY AND CONTROL OF AFLATOXIN PRODUCTION

To understand the processes involved in aflatoxin contamination of agricultural commodities, it is necessary to know the fundamental factors controlling toxin production. Aflatoxins and most other mycotoxins are secondary fungal metabolites that are not required for growth of the producing organism.[2,15,24] However, evolutionary logic requires that metabolite elaboration by specific microbes provide survival advantage to the producing species.[25] Demain[23] has provided a basis for this consideration: (1) extrinsic advantage — competitive ecological value conferred on the producing organism, and (2) intrinsic advantage — desirable metabolite factors such as shunting of excess carbon. The extrinsic advantage of a secondary metabolite as an antibiotic in an ecological niche is difficult to prove from the limited information on the natural occurrence of antibiotics. Waksman[82] warned that there was no definitive evidence linking ecological advantage to an antibiotic-producing microbe and only limited evidence for the occurrence of the substances in soil. Since aflatoxin has antibiotic properties in addition to being quite toxic and carcinogenic in primates, the widespread occurrence of the metabolite in grains and oilseeds appears to contradict the Waksman premise.[24] However, Dr. Marvin Johnson, an authority on the origin of fungal metabolites, contends that aflatoxin production in a developing crop resembles axenic culture of controlled fermentations rather than activities in an ecosystem not disturbed by humans.[103]

Dr. Johnson's observations raise important ecological questions about the role of contemporary agricultural practice and the ability of saprotrophic fungi, such as the aflatoxin-producing species, to infect developing crop plant tissues. Ecological systems can be characterized by their carbon assimilation efficiencies.[67] In a specific environment, evolutionary processes have provided for efficient capture of radiation and distribution of photosynthate. In a stable system, the intense carbon competition has provided access for a plethora of life forms. Enhanced complexity of a biological community is associated with elevated stability.[10] Abrupt environmental changes caused by natural disasters such as volcanic eruptions can upset the ecological equilibrium and provide an opportunity for colonization by species that are adapted to the new milieu. Human activities in agriculture have destabilized the natural ecosystems, to an extent linked to the scope and intensity of agricultural practices. During the past 50 years, the full impact of the new technologies has been to impose a marked homogeneity in agroecosystems. For example, the continuous cropping of limited crop species with restricted genetic variation imposes an unprecedented homogeneity, resulting in genetic vulnerability to diseases and pests. Harper[35] has summarized the situation: "Pest epidemics can often be ascribed to some activity of man that has brought floristically diverse ecosystems closer to a monoculture."

Since soil is the ultimate repository of *A. flavus* and *A. parasiticus* propagules, soil processes that influence them profoundly affect their ability to survive and establish significant inoculum levels. Generally, soil can be characterized by the absence of fermentable carbon and nitrogen compounds and microbes are found in the nonvegatative stage, e.g., conidia or sclerotia. However, the fungal species profile of a soil reflects the overlying vegetation,[21] and the observed dominance of the *Aspergillus* species can be related to certain plant materials.[80] Vegetative decay is a critical part of soil processes, so the econiche of microbial dominance is clearly dependent on the chemical composition of the substrate.

Agronomic practices such as continuous cropping of maize, and fertilization and pesticide applications, have been linked to the quantitative distribution of soil fungi, including *A. flavus*.[1,59] No *A. flavus* propagules were detected in virgin prairie soils.[5] In a study of the presence of *A. flavus* in groundnut soils, continuous cropping has been associated with increased levels of fungus.[31,32,68] *A. flavus* has also been observed as a common inhabitant of a number of maize tissues in addition to kernels.[74] With appropriate amendments, *A. flavus* can produce aflatoxin during development in the soil.[6] Presence of aflatoxin in the soil environment distinctly influenced microbial profiles.[6] In addition, the toxin can be extracted from the milieu by maize seedlings and translocated from the roots.[62] Therefore, the toxin could be recovered from soil by plant and exert a biological impact on a larger ecosystem. The soil clearly represents an opportunity to intercede and reduce the propagule potential of aflatoxin-producing fungi.[10] Introduction of varied plant residues from crop rotations would be a relatively simple method to introduce variation. Future experiments should be carried out to define the optimum conditions for increases in microbial competitors in soils or specific agronomic practices that would reduce the toxin-producing population.

Laboratory studies of the competitive ability of *A. flavus* and *A. parasicitus* in mature seeds have provided interesting but somewhat ambiguous results. Boller and Schroeder[14] examined rice cultures simultaneously inoculated with *A. parasiticus* and *A. chevalieri*. In every instance the mixed cultures produced less aflatoxin than fermentations with *A. parasiticus* alone. However, similar studies with other microbes have identified increased toxin accumulation in mixed fermentations.[3,85] Wicklow et al.[88] compared the competitive ability of *A. flavus* with other fungi on sterilized maize kernels, and found that *A. niger* and *Trichoderma viride* inhibited aflatoxin development by *A. flavus*. A similar study by Misra et al.[63] identifed the ability of *A. niger* to restrict aflatoxin production on sterilized maize kernels that were inoculated with *A. parasiticus*. The laboratory studies of microbial competition with aflatoxin-producing fungi on postharvest maize have produced provocative concepts on the nature of the interacting mycoflora on the mature seed. In addition, the investigations provide a sound basis for manipulation of the storage environment to encourage/discourage the activities of specific microbes. Preliminary observations of genotype differences in microbial selectivity in postharvest maize kernels have intriguing potential for future breeding strategies to achieve desired objectives in storage.[17,64]

Effective biological control of aflatoxin-producing fungi in developing maize kernels requires information about the range of fungal species that can effectively compete in the econiche. Studies of the fungal profiles of maize kernels have identified predominant genera of *Aspergillus*, *Penicillium*, *Fusarium*, and *Rhizopus*.[27,33,81] Although some surveys of freshly harvested kernels from the U.S. Corn Belt have observed a low incidence of *A. flavus*,[70] other studies have demonstrated incidence levels exceeding 90%.[45] High levels of *A. flavus* occurrence have been routinely observed in kernels from the southern U.S.[37,45,60] A mycological examination of kernels obtained from Corn Belt (Iowa) kernels demonstrated the predominance of three species: *A. flavus*, *Penicillium oxalicum*, and *Fusairum moniliforme*. Other studies have observed a similar restriction to a limited number of predominant species, including the three noted plus *Penicillium funiculsum* and the *Rhizopus* spp.[33,37] The limited fungal diversity on maize kernels is a startling testament to the extremely narrow range of microbial species in the contemporary agroecosystem. Prior to the era of intensive agriculture, test ears would have been infected by a number of fungal species, including isolates of *Diplodia* and *Ustilago*. Introduction of genetic resistance to the kernel-rotting pathogens has opened a unique econiche for the more opportunistic *Aspergillus* spp., including the aflatoxin producers. The widespread occurrence of *A. flavus* in developing maize kernels underscores Daubenmire's[22] prophetic statement, "For every change in one factor, a different optimum of all other factors comes into existence." To achieve desired objectives such as control of toxin-producing fungi, a comprehensive understanding of ecological aspects of the interactions among varied life forms in an agricultural context will be required.

A number of studies have identified the ecological relationships between microbes and host plants.[10] Biological control strategies that use the *Aspergillus* and *Penicillium* spp. have been effective in interfering with *Fusarium*-mediated maize seedling blight.[43] Studies of groundnut pod rots have provided convincing evidence for the competitive ability of isolates of *Trichoderma viride* in reducing *A. flavus* colonization of the pericarp, but the antagonism was abrogated by *P. funiculosum*.[86] An antagonism has also been observed between *A. niger* and *Sclerotium bataticola* against *A. flavus* in establishment of infections in groundnuts. The environmental implications of the infection of developing maize kernels by the two aflatoxin-producing species have also been investigated.[16] *A. flavus* produces exclusively aflatoxin B_1 and B_2, whereas *A. parasiticus* synthesizes all four toxins, i.e., B_1, B_2, G_1, G_2.[24] *A. flavus* routinely provides single-spore isolates that lack the ability to produce toxin, but *A. parasiticus* generally is a stable toxin producer.[26] The observations demonstrate a profound difference in the genetic characters required for toxin synthesis in the two species. The mystery of the variation is compounded by recognizing that aflatoxin-contaminated commodities contain predominantly B_1 and B_2, with a restricted occurrence of all four toxins.[27,48] To clarify the observed differences between the two species, Calvert et al.[16] inoculated developing maize kernels with varied ratios of the fungi. The ratios of aflatoxin B_1 to G_1 in mature kernels was used as a measure of the ability of the two fungal species to compete and elaborate toxin. The tests were carried out in hybrids with varied pericarp thickness. Production of the two toxins was highest in thin-pericarp kernels. The results showed that *A. parasiticus* in mixtures with *A. flavus* is severely limited in development.

VII. SUMMARY

Agricultural commodities such, as maize, being contaminated with aflatoxin caused by *Aspergillus flavus* and *A. parasiticus* has increased the research effort on this problem during the past $2\frac{1}{2}$ decades. Originally, most of the research was devoted to the stored grain aspect that included the effects of moisture, temperature, and aeration on toxin development. Later, the discovery of preharvest infection of developing maize kernels by *A. flavus* and *A. parasiticus* increased the research effort on the reduction of the contamination level and/or elimination of the toxin. The genetic control of the toxin level through host-plant resistance shows promise and should be pursued vigorously. The effect of the environment has been shown to be very important as indicated by high maize genotype by location interactions. Stress in the form of high temperatures and low plant moisture during the grain filling period enhances aflatoxin levels. Insects that damage ears during development have been shown to be vectors for the fungus. Research strategies to help solve the problem should include: (1) effect of environment, (2) development of early warning systems, (3) more information on the biology of the fungus, and (4) increased host-plant resistance for decreased toxin contamination, greater plant stress endurance, and less ear damage by insects.

Contemporary mycotoxicologists are challenged with developing appropriate concepts and procedures for managing toxin-tainted commodities to reduce negative effects in humans and domestic animals. The environmental influences responsible for selection of toxin-producing fungi are part of the current agroecosystem. The pressure for selection of toxin-producing fungi will probably increase as genetic diversity of the pertinent flora is narrowed by the commitment of more of the earth's land to agriculture. The ability to intercede and introduce procedures that block the selection of toxin-producing fungi will require extensive research information and adaptation to current technologies. Crop breeding will significantly contribute to development of remedial procedures. However, past approaches will require modification to achieve new and demanding objectives. Baker and Cook[10] succinctly summarize the situation: "Although plant breeders consider host-resistance to pathogens a form of biological control, they have concentrated on the simpler direct host-parasite effects and have rarely studied host genes that may influence interactions among microorganisms."

VIII. FUTURE STRATEGIES

Aflatoxin contamination in maize is a very complex problem that involves plant, insect, fungus, and environment. Therefore it is unlikely that any one aspect, such as resistance to the fungus, will provide a solution to the problem. Future research will need the continued cooperation of many disciplines.

The very high experimental error, plus a large genotype-by-location interaction, associated with host-plant resistance field studies, requires much more research to solve this problem. Suitable inoculation methods for field studies must be developed to provide uniform contamination by *A. flavus*. The use of the maize weevil as a vector shows promise and more research is warranted as well as a search for more vectors. Studies to reduce environmental effects in field host-plant resistance in the field would provide relevant information.

Screening for kernel resistance to *A. flavus* in the laboratory has been very limited, but the method would provide distinct advantages. Nagarajan and Bhat,[65] Wallin and Loonan,[83] Jones and Wallin,[40] and Wallin et al.[84] have used several variations of in vitro studies. In general, mature kernels were subjected to *A. flavus* spores in petri dishes, kept moist, and incubated. Methods of rating response included *A. flavus* growth, bright green yellow fluorescence, aflatoxin content, germination rates, and subsequent plant development from infected kernels. Work plans include screening for resistance to *A. flavus* infection by subjecting large numbers of mature kernels to *A. flavus* spores in vitro, incubating for 5 to 8 days, propagating the kernels that germinate, and producing mature plants and ears. Several cycles of repetition of the process may well be necessary to determine if the method is successful. Wallin has completed two cycles of the procedure, and the tentative results are encouraging.[104]

More research is needed on the role of trace elements such as boron, copper, manganese, molybdenum, and zinc in *A. flavus* growth and aflatoxin synthesis during kernel development. If some of the elements are required by the fungus, various maize genotypes could be examined to determine if differences for trace element levels occur in developing kernels. The effect on the minor element content in the grain produced by adding trace elements to the soil should also be examined. Hinsley et al.[39] found large differences among 20 maize inbred lines for Zn and Cd accumulation in leaves and grain when the plants were grown in soil amended with sewage sludge. The authors concluded that accumulation of these two elements may be under host-plant genetic control.

The relationships between toxin-producing and nontoxin-producing isolates of *A. flavus* need clarification. The role of toxin production in the infection process is not well understood; it could represent an important part of the selection process. Nutritional requirements for optimum *A. flavus* growth should be established, for example, by varying the media with levels of amino acids, fatty acids, starch, and other substances that vary sufficiently in the kernels to cause varying levels in aflatoxin. Greater pericarp thickness and quality may provide a barrier to *A. flavus* infection. More information on *A. flavus* spore germination and growth on silks and in cobs would be useful.

Also, more research is needed to determine the effect of stress on aflatoxin contamination. Types of stress, such as drought, high temperatures, soil moisture, and predation should be examined through techniques such as root pruning, varied degrees of shading, leaf removal at various stages of growth, and simulations of aggressive insect attack.

REFERENCES

1. **Abdel-Kader, M. I. A., Moubasher, A. H., and Abdel-Hafez, S. I.,** Selective effects of five pesticides on soil and cotton rhizosphere-and-rhizoplane fungus flora, *Mycopathologia*, 66, 117, 1978.
2. **Aharonowitz, Y. and Demain, A. L.,** Thoughts on secondary metabolism, *Biotechnol. Bioeng.*, 22, 5, 1980.
3. **Alderman, G. G., Emeh, C. O., and Marth, E. H.,** Aflatoxin and rubratoxin produced by *Aspergillus parasiticus* and *Penicillium rubrum* when grown independently, associately or with *Penicillium italicum* or *Lactobacillus plantarum*, *Z. Lebensm. Unters. Forsch.*, 153, 305, 1973.
4. **Anderson, H. W., Nehring, E. W., and Wicher, W. R.,** Aflatoxin contamination of corn in the field, *J. Agric. Food Chem.*, 23, 775, 1975.
5. **Angle, J. S., Dunn, K. A., and Wagner, G. H.,** Effect of cultural practices on the soil population of *Aspergillus flavus* and *Aspergillus parasiticus*, *Soil Sci. Am. J.*, 46, 301, 1982.
6. **Angle, J. S. and Wagner, G. H.,** Aflatoxin B₁ effects on soil microorganisms, *Soil Biol. Biochem.*, 13, 381, 1981.

7. **Ashworth, L. J., Schroeder, H. W., and Langley, B. C.,** Aflatoxins: environmental factors governing occurrence in Spanish peanuts, *Science,* 148, 1228, 1965.

8. **Asplin, F. D. and Carnaghan, R. B. A.,** The toxicity of certain ground nut meals for poultry with special reference to their effect on ducklings and chickens, *Vet. Rec.,* 73, 1215, 1961.

9. **Ayerst, G.,** Effect of moisture and temperature on growth and spore germination in some fungi, *J. Stored Prod. Res.,* 5, 127, 1969.

10. **Baker, K. F. and Cook, R. J.,** *Biological Control of Plant Pathogens,* W. H. Freeman & Co., San Francisco, 1974.

11. **Barry, D., Zuber, M. S., Lillehoj, E. B., McMillian, W. W., Adams, N. J., Kwolek, W. F., and Widstrom, N. W.,** Evaluation of two arthropod vectors as inoculators of developing maize ears with *Aspergillus flavus, Environ. Entomol.,* 14, 634, 1985.

12. **Barry, D., Lillehoj, E. B., Widstrom, N. W., McMillian, W. W., Zuber, M. S., Kwolek, W. F., and Guthrie, W. D.,** Effect of husk tightness and insect infestation on aflatoxin contamination of preharvest maize, *Environ. Entomol.,* 15, 1116, 1986.

13. **Blount, W. P.,** Turkey "x" diseases, *Turkeys (J. Brit. Turkey Federation),* 9(2), 52, 1961.

14. **Boller, R. A. and Schroeder, H. W.,** Influence of *Aspergillus chevalieri* on production of aflatoxin in rice by *Aspergillus parasiticus, Phytopathology,* 64, 17, 1973.

15. **Bu'Lock, J. D.,** *The Biosynthesis of Natural Products,* McGraw-Hill, New York, 1965.

16. **Calvert, O. H., Lillehoj, E. B., Kwolek, W. F., and Zuber, M. S.,** Aflatoxin B_1 and G_1 production in developing *Zea mays* kernels from mixed inocula of *Aspergillus flavus* and *A. parasiticus, Phytopathology,* 68, 501, 1978.

17. **Cantone, F. A., Tuite, J., Bauman, L. F., and Stroshine, R.,** Genotypic differences in reaction of stored corn kernels to attack by selected *Aspergillus flavus* and *Penicillum* spp., *Phytopathology,* 73, 1250, 1983.

18. **Christensen, C. M.,** Deterioration of stored grains by fungi, *Bot. Rev.,* 23, 108, 1957.

19. **Christensen, C. M. and Kaufman, H. H.,** Grain storage, in *The Role of Fungi in Quality Loss,* University of Minnesota Press, Minneapolis, Minn., 1969.

20. **Christensen, M.,** A synoptic key and evaluation of species in the *Aspergillus flavus* group, *Mycologia,* 73, 1056, 1981.

21. **Christensen, M.,** The soil microfungi of conifer hardwood forests in Wisconsin, Ph.D. dissertation, University of Wisconsin, Madison, Wis., 1960.

22. **Daubenmire, R.,** *Plant and Environment: A Textbook of Autecology,* 2nd ed., John Wiley & Sons, New York, 1959.

23. **Demain, A.,** How do antibiotic-producing microorganisms avoid suicide, *Ann. New York Acad. Sci.,* 235, 601, 1974.

24. **Detroy, R. W., Lillehoj, E. B., and Ciegler, A.,** Aflatoxin and related compounds, in *Microbiol Toxins,* Vol. 6, Ciegler, A., Kadis, S., and Ajl, S. J., Eds., Academic Press, New York, 1971, 3.

25. **Dobzhansky, T.,** *Genetics of the Evolutionary Process,* Columbia Press, New York, 1970.

26. **Fennell, D. I., Bothast, R. J., Lillehoj, E. B., and Peterson, R. E.,** Bright greenish-yellow fluorescence and associated fungi in white corn naturally contaminated with aflatoxin, *Cereal Chem.,* 50, 404, 1973.

27. **Fennell, D. I., Lillehoj, E. B., and Kwolek, W. F.,** *Aspergillus flavus* and other fungi associated with insect-damaged field corn, *Cereal Chem.,* 52, 314, 1975.

28. **Fennell, D. I., Lillehoj, E. B., Kwolek, W. F., Gutherie, W. D., Sheely, R., Sparks, A. N., Widstrom, N. W., and Adams, G. L.,** Insect larval activity on developing corn ears and subsequent aflatoxin contamination of the seed, *J. Econ. Entomol.,* 71, 624, 1978.

29. **Gardner, C. A. C., Darrah, L. L., Zuber, M. S., and Wallin, J. R.,** Genetic control of aflatoxin production in maize, Ph.D. thesis of senior author, University of Missouri, Columbia, Mo., 1986.

30. **Gray, F. A., Faw, W. F., and Boutwell, J. L.,** The 1977 corn-aflatoxin epiphytotic in Alabama, *Plant Dis.,* 66, 221, 1982.

31. **Griffin, G. J. and Garren, K. H.,** Colonization of rye green manure and peanut fruit debris by *Aspergillus flavus* group in field soils, *Appl. Environ. Microbiol.,* 32, 28, 1976.

32. **Griffin, G. J., Garren, K. H., and Taylor, J. D.,** Influence of crop rotation and minimum tillage on the population of *Aspergillus flavus* group in peanut field soil, *Plant Dis.,* 65, 898, 1981.

33. **Gulya, T. J., Martinson, C. A., and Tiffany, L. H.,** Ear-rotting fungi associated with opaque-2 maize, *Plant Dis. Rep.,* 63, 370, 1979.

34. **Hart, J. R.,** A method for detecting mixtures of artificial dried corn and high moisture, *Cereal Chem.,* 44, 601, 1967.

35. **Harper, J. L.,** *Population Biology of Plants,* Academic Press, New York, 1977.

36. **Hesseltine, C. W., Sorenson, W. G., and Smith, M.,** Taxonomic studies of the aflatoxin-producing strains in the *Aspergillus flavus* group, *Mycologia,* 62, 123, 1970.

37. **Hesseltine, C. W., Shotwell, O. L., Kwolek, W. F., Lillehoj, E. B., Jackson, W. K., and Bothast, R. J.,** Aflatoxin occurrence in 1973 corn at harvest. II. Mycological studies, *Mycologia,* 68, 341, 1976.

38. **Hesseltine, C. W., Rogers, R. F., and Shotwell, O. L.,** Aflatoxin and mold flora in North Carolina 1977 corn crop, *Mycologia,* 73, 216, 1981.

39. **Hinesly, T. D., Alexander, D. E., Ziegler, E. L., and Barrett, G. L.,** Zinc and Cd accumulation by corn inbreds grown on sludge amended soil, *Agron. J.,* 70, 425, 1978.

40. **Jones, D. L. and Wallin, J. R.,** Evaluation of exotic corn for resistance to aflatoxin formation by *Aspergillus flavus, Phytopathol. News,* 12, 141, 1978.

41. **Jones, R. K., Duncan, H. E., Payne, G. A., and Leonard, K. J.,** Factors influencing infection by *Aspergillus flavus* in silk-inoculated corn, *Plant Dis.,* 64, 859, 1980.

42. **Koehler, B.,** Fungus growth in shelled corn as affected by moisture, *J. Agric. Res.,* 56, 291, 1938.

43. **Kommedahl, T.,** Utilization of biological agents other than host resistance for control of plant pathogens, in *Proc. Inst. Biol. Control Plant Insects Disease,* Maxwell, F. G. and Harris, F. A., Eds., University Press Mississippi, Mississippi State University, State College, Miss., 1974, 248.

44. **Lillehoj, E. B., Kwolek, W. F., Shannon, G. M., Shotwell, O. L., and Hesseltine, C. W.,** Aflatoxin occurrence in 1973 field corn. I. A limited survey in the southeastern U.S., *Cereal Chem.,* 52, 603, 1975.

45. **Lillehoj, E. B., Kwolek, W. F., Manwiller, A., DuRant, J. A., LaPrade, J. C., Horner, E. S., Reid, J., and Zuber, M. S.,** Aflatoxin production in several corn hybrids grown in South Carolina and Florida, *Crop Sci.,* 16, 483, 1976.

46. **Lillehoj, E. B., Fennell, D. I., and Hesseltine, C. W.,** *Aspergillus flavus* infection and aflatoxin production in mixture of high-moisture and dry maize, *J. Stored Prod. Res.,* 12, 11, 1976.

47. **Lillehoj, E. B., Fennell, D. I., and Kwolek, W. F.,** Aflatoxin and *Aspergillus flavus* occurrence in 1975 corn at harvest from a limited region of Iowa, *Cereal Chem.,* 54, 366, 1976.

48. **Lillehoj, E. B., Fennell, D. I., and Kwolek, W. F.,** *Aspergillus flavus* and aflatoxin in Iowa corn before harvest, *Science,* 193, 495, 1976.

49. **Lillehoj, E. B., Kwolek, W. F., Zuber, M. S., Calvert, O. H., Horner, E. S., Widstrom, N. W., Guthrie, W. D., Scott, G. E., Thompson, D. L., Findley, W. R., and Bockholt, A. J.,** Aflatoxin contamination of field corn: Evaluation of regional test plots for early detection, *Cereal Chem.,* 55, 1007, 1978.

50. **Lillehoj, E. B., Kwolek, W. F., Horner, E. S., Widstrom, N. W., Josephson, L. M., Franz, A. O., and Catalano, E. A.,** Aflatoxin contamination of preharvest corn: Role of *Aspergillus flavus* inoculum and insect damage, *Cereal Chem.,* 57, 255, 1980.

51. **Lillehoj, E. B., Kwolek, W. F., Zuber, M. S., Bockholt, A. J., Calvert, O. H., Findley, W. R., Guthrie, W. D., Horner, E. S., Josephson, L. M., King, S., Manwiller, A., Sauer, D. B., Thompson, D., Turner, M., and Widstrom, N. W.,** Aflatoxin in corn before harvest: Interaction of hybrids and locations, *Crop Sci.,* 20, 731, 1980.

52. **Lillehoj, E. B., Kwolek, W. F., Zuber, M. S., Horner, E. S., Widstrom, N. W., Gutherie, W. D., Turner, M., Sauer, D. B., Findley, W. R., Manwiller, A., and Josephson, L. M.,** Aflatoxin contamination caused by natural fungal infection of preharvest corn, *Plant Soil,* 54, 469, 1980.

53. **Lillehoj, E. B.,** Effect of environmental and cultural factors on aflatoxin contamination of developing corn kernels, in Aflatoxin and *Aspergillus flavus* in corn, *South. Coop. Ser. Bull.* 279, Diener, U. L., Asquith, R. L., and Dickens, J. W., Eds., Auburn University, Auburn, Ala., 1983, 27.

54. **Lillehoj, E. B., Zuber, M. S., Darrah, L. L., Kwolek, W. F., Findley, W. R., Horner, E. S., Scott, G. E., Manwiller, A., Sauer, D. B., Thompson, D., Warren, H., West, D. R., and Widstrom, N. W.,** Aflatoxin occurrence and levels in preharvest corn kernels with varied endosperm characteristics grown at diverse locations, *Crop Sci.,* 23, 1181, 1983.

55. **Lin, Y. N., Ayres, J. C., and Koehler, P. E.,** Influence of temperatures cycling on the production of aflatoxin B_1 and G_1 by *Aspergillus parasiticus, Appl. Environ, Microbiol.,* 40, 333, 1980.

56. **Loosmore, R. M. and Harding, J. D. J.,** A toxic factor in Brazilian groundnut causing liver damage in pigs, *Vet. Rec.,* 73, 1362, 1961.

57. **Lopez, L. C. and Christensen, C. M.,** Effect of moisture content and temperature on invasion of stored corn by *Aspergillus flavus, Phytopathology,* 57, 588, 1967.

58. **Maing, I.-Y., Ayres, Y. C., and Koehler, P. E.,** Persistence of aflatoxin during fermentation of soy sauce, *Appl. Microbiol.,* 25, 1015, 1973.

59. **Martyniuk, S. and Wagner, G. H.,** Quantitative and qualitative examination of soil microflora associated with different management systems, *Soil Sci.,* 125, 343, 1978.

60. **McMillian, W. W., Wilson, D. M., and Widstrom, N. W.,** Insect damage, *Aspergillus flavus* ear mold and aflatoxin contamination in South Georgia corn fields in 1977, *J. Environ. Qual.,* 7, 564, 1978.

61. **McMillian, W. W., Wilson, D. M., and Widstrom, N. W.,** Aflatoxin contamination of preharvest corn in Georgia: A six-year study of insect damage and visible *Aspergillus flavus, J. Environ. Qual.,* 14, 200, 1985.

62. **Mertz, D., Lee, D., Zuber, M., and Lillehoj, E.,** Uptake and metabolism of aflatoxin by *Zea mays, J. Agric. Food Chem.,* 28, 963, 1980.

63. **Misra, R. S., Sinha, K. K., and Sinha, P.,** Aflatoxin production by *Aspergillus parasiticus* (NRRL-3240) on maize seeds in competitive environment, *Nat. Acad. Sci. Lett.,* 4, 123, 1981.

64. **Moreno-Martinez, E. and Christensen, C. M.,** Differences among lines and varieties of maize in susceptibility to damage by storage fungi, *Phytopathology,* 61, 1498, 1971.

65. **Nagarajan, V. and Bhat, R. V.,** Factors responsible for varietal differences in aflatoxin production in maize, *J. Agric. Food Chem.,* 20, 911, 1972.

66. **Northolt, M. D. and Bullerman, L. B.,** Prevention of mold growth and toxin production through control of environmental conditions, *J. Food Protect.,* 45, 519, 1982.

67. **Odum, H. T. and Odum, E. C.,** *Energy Basis for Man and Nature,* McGraw-Hill, New York, 1976.

68. **Petit, R. E. and Taber, R. A.,** Factors influencing aflatoxin accumulation in peanut kernels and the associated mycoflora, *Appl. Microbiol.,* 16, 1230, 1968.

69. **Quasem, S. A. and Christensen, C. M.,** Influence of moisture content, temperature and time on the deterioration of stored corn by fungi, *Phytopathology,* 48, 544, 1958.

70. **Rambo, G. W., Tuite, J., and Caldwell, R. W.,** *Aspergillus flavus* in preharvest corn from Indiana in 1971 and 1972, *Cereal Chem.,* 51, 595, 1974.

71. **Sauer, D. B. and Burroughs, R.,** Fungal growth, aflatoxin production and moisture equilibration in mixtures of wet and dry corn, *Phytopathology,* 70, 516, 1980.

72. **Schroeder, H. W. and Hein, H., Jr.,** Effect of diurnal temperature cycles on the production of aflatoxin, *Appl. Microbiol.,* 16, 988, 1968.

73. **Shotwell, O. L.,** Aflatoxin in corn, *J. Am. Oil Chem. Soc.,* 54, 216A, 1977.

74. **Shotwell, O. L., Goulden, M. L., Hesseltine, C.W., Dickens, J. W., and Kwolek, W. F.,** Aflatoxin distribution in contaminated corn plants, *Cereal Chem.,* 57, 206, 1980.

75. **Stutz, H. K. and Krumperman, P. H.,** Effect of temperature cycling on the production of aflatoxin by *Aspergillus parasiticus, Appl. Environ. Microbiol.,* 32, 327, 1976.

76. **Taubenhaus, J. J.,** A study of the black and yellow molds of ear corn, *Texas Agric. Exp. Sta. Bull.,* 270, 38, 1920.

77. **Thompson, D. L., Lillehoj, E. B., Leonard, K. J., Kwolek, W. F., and Zuber, M. S.,** Aflatoxin concentraction in corn as influenced by kernel development state and post-inoculation temperature in controlled environments, *Crop Sci.,* 20, 609, 1980.

78. **Thompson, R. A. and Foster, G. H.,** Stress cracks in artificially dried corn, *U. S. Dept. of Agric. Mark. Res. Rep.,* 631, 1, 1963.

79. **Trenk, H. L. and Hartman, P. A.,** Effects of moisture content and temperature on aflatoxin production in corn, *Appl. Microbiol.,* 19, 781, 1970.

80. **Upadhyay, R. E. and Rai, B.,** Ecological survey of Indian soil fungi with special reference to Aspergilli, Penicillia, and Trichoderma, *Rev. Ecol. Biol. Sci.,* 16, 39, 1979.

81. **Vaidehi, B. K. and Ramarao, P.,** Fungi from maize seed, *Natl. Acad. Sci. Lett.,* 1, 283, 1978.

82. **Waksman, S.,** Antibiotics, *Biol. Rev.,* 23, 452, 1948.

83. **Wallin, J. R. and Loonan, D. V.,** A method of trapping and identifying spores of *Aspergillus flavus, Plant Dis. Rep.,* 60, 918, 1976.

84. **Wallin, J. R., Zuber, M. S., Loonan, D. V., and Gardner, C. A. C.,** The response of inbred lines of maize challenged with *Aspergillus flavus, Trans. Mo. Acad. Sci.,* 14 (Abstr.), 174, 1980.

85. **Weckbach, L. W. and Marth, E. H.,** Aflatoxin production in a competitive environment, *Mycopathologia,* 62, 39, 1977.

86. **Wells, T. R., Kreutzer, W. A., and Lindsey, D. L.,** Colonization of gnotobiotically grown peanuts by *Aspergillus flavus* and selected interacting fungi, *Phytopathology,* 62, 1238, 1972.

87. **West, S., Wyatt, R. D., and Hamilton, P. B.,** Improved yield of aflatoxin by incremental increases of temperature, *Appl. Microbiol.,* 25, 1018, 1973.

88. **Wicklow, D. T., Hesseltine, C. W., Shotwell, O. L., and Adams, G. L.,** Interference competition and aflatoxin levels in corn, *Phytopathology,* 70, 761, 1980.

89. **Widstrom, N. W., Sparks, A. N., Lillehoj, E. B., and Kwolek, W. F.,** Aflatoxin production and lepidopteran insect injury on corn in Georgia, *J. Econ. Entomol.,* 68, 855, 1975.

90. **Widstrom, N. W., Wiseman, B. R., McMillian, W. W., Kwolek, W. F., Lillehoj, E. B., Jellum, M. D., and Massey, J. H.,** Evaluation of commercial and experimental three-way corn hybrids for aflatoxin B_1 production potential, *Agron. J.,* 70, 986, 1978.

91. **Widstrom, N. W.,** The role of insects and other plant pests in aflatoxin contamination of corn, cotton and peanuts. A review, *J. Environ. Qual.,* 8, 5, 1979.

92. **Widstrom, N. W. and Zuber, M. S.,** Prevention and control of aflatoxin in corn: Sources and mechanisms of genetic control in the plant, in Aflatoxin and *Aspergillus flavus* in Corn, *South. Coop. Ser. Bull.* 279, Diener, U. L., Asquith, R. L., and Dickens, J. W., Eds., Auburn University, Auburn, Ala., 1983, 72.

93. **Widstrom, N. W., Wilson, D. M., and McMillian, W. W.,** Ear resistance of maize inbreds to field aflatoxin contamination, *Crop Sci.,* 24, 1155, 1984.

94. **Wilson, D. M. and Jay, E.,** Influence of modified atmosphere storage on aflatoxin production in high-moisture corn, *Appl. Microbiol.,* 29, 224, 1975.

95. **Wilson, D. M., McMillian, W. M., and Widstrom, N. W.,** Field aflatoxin contamination of corn in south Georgia, *J. Am. Oil Chem. Soc.,* 56, 798, 1979.

96. **Zuber, M. S., Calvert, O. H., Lillehoj, E. B., and Kwolek, W. F.,** Preharvest development of aflatoxin B₁ in corn in the United States, *Phytopathology,* 66, 1120, 1976.

97. **Zuber, M. S.,** Influence of plant genetics on toxin production in corn, in *Mycotoxins in Human and Animal Health,* Rodricks, J. V., Hesseltine, C. W., and Mehlman, M. A., Eds., Pathotox Publishers, Park Forest South, Ill., 807.

98. **Zuber, M. S., Calvert, O. H., Kwolek, W. F., Lillehoj, E. B., and Kang, M. S.,** Aflatoxin B₁ production in an eight-line diallele of *Zea mays* infected with *Aspergillus flavus, Phytopathology,* 68, 1346, 1978.

99. **Zuber, M. S. and Lillehoj, E. B.,** Status of the aflatoxin problem in corn, *J. Environ. Qual.,* 8, 1, 1979.

100. **Zuber, M. S., Darrah, L. L., Lillehoj, E. B., Josephson, L. M., Manwiller, A., Scott, G. E., Gudauskas, R. T., Horner, E. S., Widstrom, N. W., Thompson, D. L., Bockholt, A. J., and Brewbaker, J. L.,** Comparison of open-pollinated maize varieties and hybrids for preharvest aflatoxin contamination in the southern United States, *Plant Dis.,* 67, 185, 1983.

101. **Sauer, D.,** personal communication.

102. **Gardner, C. A. C.,** et al., unpublished data.

103. **Johnson, M.,** personal communication.

104. **Wallin, J. R.,** personal communication.

Chapter 8

DUTCH ELM DISEASE, A MODEL TREE DISEASE FOR BIOLOGICAL CONTROL*

Rudy J. Scheffer and Gary A. Strobel

TABLE OF CONTENTS

* Rudy J. Scheffer gratefully acknowledges a grant from the Netherlands Organization for the Advancement of Pure Research (Z.W.O.) supporting this work.

I. INTRODUCTION

The elm has been used in the old world throughout the ages because of its unique combination of wind and salt tolerance, fast growth on almost every soil, valuable timber, and its ability to withstand transplanting, lopping, and root pruning. In ancient times, foliage and bark were used for fodder or even for human nutrition. Later, elm became an important timber, and in medieval Europe it was even second in importance.[36,37,68,98]

Elm became the most widely planted tree throughout much of its potential habitat, especially in large parts of Europe and, later, North America. It has become one of the best known of all tree genera, and town roads and buildings are named after it all over Europe and America.

The constant interest in elm made it a common tree, planted for timber, shade, beauty, or shelter, but also made it *the* typical nonwoodland tree. As such, only a limited gene pool was used in planting, usually even monospecies stands. This turned out to be the prelude to disaster. *The* elm disease, once established, could thus spread across continents, killing trees by the millions.

Currently, Dutch elm disease is the most widely known plant disease in the western world. It also is the most intensively studied tree disease and a prime example of a disease for which virtually every approach for control has been pursued. Recent results on control warrant this review.

The story of Dutch elm disease thus represents a model for vascular diseases, as the life cycle of the pathogen is known in ever increasing detail, and transmittance of the disease is by vectors or root contact. Also, artificial infection of the host is simple, and trees having various degrees of resistance are known.

II. HISTORY OF DUTCH ELM DISEASE

In 1919 in the southern Netherlands, elms were discovered showing symptoms of a yet unknown disease: a sudden wilting and dying of the leaves and branches. Trees, which in early summer appeared normal, in full leaf, sometimes withered, lost all their leaves and died within a matter of weeks. In others, the leaves on a few branches in the crown turned yellow and fell, and by late summer these symptoms spread over the crown with no distinct boundaries. On shoots overtaken by the disease during growth, the end leaves on the withered and stunted tips often remained after the fully grown leaves had fallen off, thereby producing a characteristic "shepherd's crook" effect. In addition, diseased branches, when cut, always revealed a dark discoloration of the wood, at any rate, in the most recent growth ring. The disease was described by Spierenburg.[83,84] At the Willie Commelin Scholten Phytopathological Laboratory, Schwarz[79] isolated one fungus from discolored current-year sapwood that never grew from healthy elm wood. On the basis of the coremia formed she described the fungus as *Graphium ulmi,* and she concluded that this was the pathogen causing the disease in the elms. However, others held different opinions, from bacteria to climatical or soil factors, or even mustard gas that was used in the great war.[21,39,40] Wollenweber[100] and Westerdijk[96] proved that *G. ulmi* was indeed the causal organism.

Wollenweber[101] found *G. ulmi* in diseased elms in galleries of the larger elm bark beetle, *Scolytus scolytus;* and Betrem[6] isolated the fungus from this beetle. The following papers of Roepke,[71] Fransen,[28-30] and Fransen and Buisman[31] made it clear that the elm bark beetles *S. scolytus* and *S. multistriatus* were vectors of *G. ulmi.*

After the discovery of the sexual stage of the pathogen[13] the name was changed into *Ceratostomella ulmi* (Schwarz) Buisman. Later, Nannfeldt classified the fungus as *Ophiostoma ulmi* (Buisman) Nannf.[57] Moreau,[59] Hunt,[45] and Upadhyay[91] proposed *Ceratocystis ulmi* (Buisman) C. Moreau. However, on the basis of chemotaxonomical and anamorph characters, the name *Ophiostoma ulmi* should be preferred.[43,97] The discovery of Dutch elm disease in elms considered to be resistant led to distinguishing a nonaggressive and an aggressive strain,[92,93] which probably represent subspecies of *O. ulmi*. Within the aggressive strain a North-American (NAN) race and an Eurasian (EAN) race are distinguishable.[9,10]

The elm disease, which, due to the significant research efforts of Dutch scientists, became known as the Dutch elm disease, was discovered in the southern Netherlands in 1919, as mentioned above. However, Guyot[33] described a disease which killed elms in Bussy in northern France in 1918 with symptoms resembling Dutch elm disease. We may assume the disease to have existed, unnoticed, even several years earlier in the war-stricken areas. After 1919 Dutch elm disease rapidly became widespread. In 1923 it occurred in England, although it was not identified until 1927.[99] The first confirmation of the disease in North America came from Cleveland, Ohio, in 1930.[12,55] Most probably, the fungus was imported with veneer logs from France.[1] One of the potential vectors, the smaller (European) elm bark beetle *S. multistriatus,* was already known on this continent since 1909.[15] Also, an elm bark beetle native to North America, *Hylurgopinus rufipes,* proved to be an effective vector of *O. ulmi*.

The major elm taxa planted in both Europe and North America proved to be susceptible to *O. ulmi,* with the exception of the majority of the Asiatic species.[81] Probably this prerequisite, and the breeding behavior of the elm bark beetles (breeding only in weakened, diseased, or recently dead elms), explain the unsurpassed devastation of the elm populations.

III. PATHOGENESIS

O. ulmi is confined to the xylem vessels of the elm at least during the initial stages of the disease.[25,58,66,90] Therefore, infection means introduction of the pathogen into a xylem vessel, usually from a feeding groove of an elm bark beetle, or via xylem anastomoses which frequently occur between roots of adjacent elms. In the xylem vessel, conidia may be formed which are carried with the sap. The pathogen spreads from one vessel to another by hyphal penetration of pit membranes (Figure 1).

Wilting of the infected elm is apparently a result of interactions between fungal metabolites and the tree; blockage of the xylem vessels by the fungus itself is most unlikely because of the usually low density of hyphae and conidia in the vessels.[19,58] Much research has been done over the years on fungal metabolites produced in vitro. Of those compounds which may interact with the tree, in particular, "toxins" and cell wall-degrading enzymes have been investigated. Several phytotoxic compounds have been isolated and purified from shake cultures of *O. ulmi*. These compounds can be classified into four groups: phenolics, polysaccharides, a protein named cerato-ulmin, and a glycopeptide.[72] However, evidence for a role in pathogenesis exists only for cerato-ulmin[89] and the glycopeptide.[4,5,76]

In vitro production of pectic and cellulolytic enzymes has been reported by several workers.[3,7,22,42,46,87] Woods and Holmes[102] were able to detect significant amounts of pectic enzyme activity in fluid expressed from infected elm wood, but only insignificant amounts from healthy trees. Only in two cases enzyme activities of nonaggressive and aggressive isolates were compared. In one, a correlation with aggressiveness was sought, but not found, for polygalacturonase and β-glucosidase (C_x) activities.[22] In the other, Svaldi and Elgersma[87] found a clear correlation comparing glycosidase and exoglycanase activities of nonaggressive and aggressive isolates of *O. ulmi:* significantly more arabinose, xylose, and rhamnose were released from cell walls of elm wood by enzyme preparations of aggressive isolates.

The extensive genotypical differences between the nonaggressive and the aggressive strain make it predictable though, that other, not yet elucidated factors play a role in the extended host range of the aggressive strain.

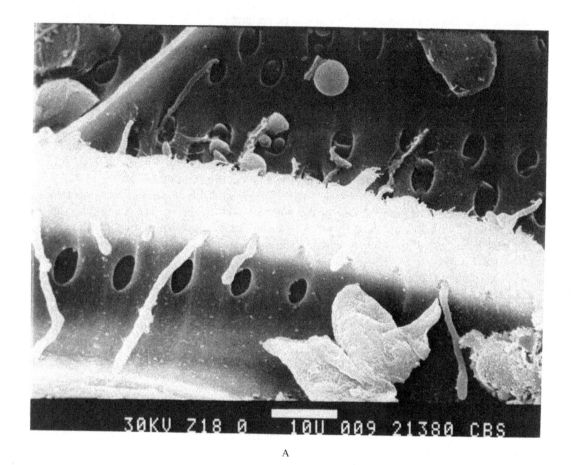

A

FIGURE 1A and 1B.   Spread of *O. ulmi* from one xylem vessel to another by penetration of pit membranes by hyphae. The xylem is the exclusive environment for the pathogen, once established within the tree.

IV. CONTROL

Various measures can be taken to control Dutch elm disease. Each one aims at a seemingly vulnerable spot in the complex relationship existing between the fungus, the beetles, and the susceptible elm species. Quarantines can theoretically exclude Dutch elm disease from an area, provided that there are elm-free natural barriers. Although both susceptible elms and elm bark beetles are present in Australia, Dutch elm disease has not yet occurred there, showing the feasibility of this approach. However, once the disease is established, eradication of the disease focus to prevent spread to other areas is virtually impossible. Vector suppression, preservation of valuable elms by injecting systemic fungicides, prevention of root graft transmission of the fungus, planting of elms with a high degree of resistance, or combinations of these are the remaining tactics that, over the years, have proven to be of some use.[81]

Considerable numbers of elms acquire Dutch elm disease by tree-to-tree transport via root grafts. When the disease starts in the root system, the curative treatments are usually ineffective. Especially if diseased elms are left standing long enough for the fungus to spread into the root system, infection of neighboring trees via root grafts is a common phenomenon.

Prevention of root grafts by digging trenches in between elms, or by killing all the roots halfway between trees by injecting a soil sterilant such as metam-sodium into the soil, is sometimes advised.[63] But, the high cost of such treatments prevents its common use, not

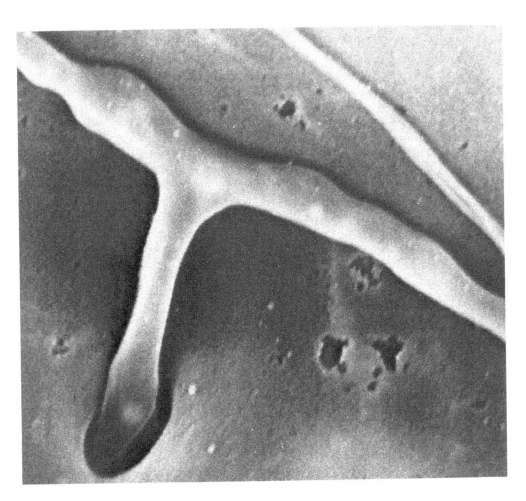

FIGURE 1B.

mentioning the technical problems of digging trenches between elms along a road with a variety of buried utility lines. An especially sad way of destroying root contact is first taking down the apparently healthy elm next to a diseased one. However, it is much better than losing the last few trees on a row year after year by ongoing root contact-mediated infections.

In the long run, replacing elms with elms resistant to the disease may be promising. Resistance of elms to Dutch elm disease is polygenetically controlled and probably quantitative in nature.[38,54] Research on factors determining susceptibility or resistance has been a major topic in Dutch elm disease research over the years, leading to an overwhelming amount of detailed information on differences between susceptible and more resistant elms. Examples are (1) faster tylose formation in resistant elms, (2) smaller diameter of xylem vessels, (3) smaller groups of xylem vessels more widely spaced, or (4) more summer wood, factors that all may contribute to hampering spread of the pathogen.[8,23,24]

However, the various breeding programs up until now have not put emphasis on these aspects, but used the practical approach of inoculating young seedlings with the pathogen. Those that survived were tested further and used for crosses in the hope of obtaining a resistant elm with a desirable shape and good growth.

Breeding for resistance started in the Netherlands at the Willie Commelin Scholten Phytopathological Laboratory in 1928. One of the original elms collected as a potential parent in the breeding program was so resistant that it was released as a clonal tree in 1936. It was named after one of the founders of the breeding program, Christine Buisman, who died untimely in 1936. In the Netherlands, this elm proved to be very prone to *Nectria cinnabarina*. The cankers often distorted the tree's growth so much that they had to be removed, thus, few of them can be found nowadays along Dutch streets. The Bea Schwarz elm was the second elm to be released from the Dutch breeding program (1947). In 1961, after Hans Heybroek had taken over the breeding program, the first hybrid clone was released: the Commelin elm. In the Netherlands it became so popular that it was used in vast quantities to replace many of the elms lost to the Dutch elm disease epidemic. It was followed in 1963 by the Groeneveld elm which was even more resistant. The breeding program is still ongoing, mainly because of the discovery of the aggressive strain of the Dutch elm disease fungus. In the Netherlands in 1972 a Commelin elm, thought to be resistant, showed Dutch elm disease symptoms. Since then many Commelins have died, but the breeding program has released four new clones, all given botanists' names: Plantyn, Lobel, Dodoens, and Clusius.

In the U.S., Donald Welch of Cornell University started collecting American elm trees in 1939. He inoculated thousands of trees in order to find some resistance. He found some, but many of the survivors died of elm phloem necrosis (or elm yellows), a disease caused by a mycoplasma-like organism. Others, like Francis Holmes (Amherst, Mass.), Larry Schreiber and Alden Townsend (Delaware, Ohio), and Gene Smalley (Madison, Wis.) carried on the search, with the result that several elms were released that are considered to be resistant. An overview is given in Table 1.

V. CONTROL OF THE VECTORS

Once established, the suppression of insect vector populations has usually been the most successful method to control Dutch elm disease. DDT, which effectively kills the beetles, was widely applied in North America and saved hundred of thousands of elms until this insecticide was banned for environmental reasons. Its successors, methoxychlor and chlorpyrifos are less effective.[52]

A natural control of elm bark beetle populations by the fungus *Phomopsis oblonga* was observed in Wales.[94] This fungus apparently makes dying or recently dead elms unattractive for breeding by elm bark beetles. If their eggs were laid in *Phomopsis*-colonized trees only a few developed into a new generation of beetles. However, massive colonization of elm bark by *Phomopsis* has mainly been observed in *Ulmus glabra,* and it is not clear if manipulation of *Phomopsis* can result in the extension of its beetle-controlling attribute to other elms or areas.

Sanitation, which must involve the prompt removal of all weakened or diseased elms, is now the basic and most effective means of vector suppression. In the Netherlands the sanitation program has been highly effective, suppressing losses to approximately 1%/year, albeit at an annual cost of more than 2^1/_2$ million paid by the central government, and a comparable sum funded by municipalities and other authorities.

VI. CONTROL OF THE PATHOGEN

Individual elms sometimes can be saved by injection of a systemic fungicide, whether or not combined with eradicative pruning or attempts to prevent root graft transmission.[14,26] The benzimidazole derivative carbendazim, sold under the trade name Lignasan, was widely applied until its ultimate usefulness came into question. Thiabendazole hypophosphite, marketed as Arbotect 20-S, Lirotect ulmi 20-S, or Ceratotect, seems to be more effective as it is retained much longer in the tree.[32] The infection may be completely suppressed, particularly if injected in a high dose, 3 g of active compound per liter, and in an amount of 1.9 ℓ/cm tree diameter at breast height (DBH), provided the wilted portions of the tree are subsequently

Table 1
ELM CULTIVARS CURRENTLY TRADED WITH A CLAIM FOR
RESISTANCE TO DUTCH ELM DISEASE.

Name	Year of release	Botanical background	Source
'Groeneveld'	1963	*U.* × *hollandica*	H. M. Heybroek
'Dodoens'	1975	Complex hybrid	Institute for Forestry and Landscape Planning
'Lobel'	1975	Complex hybrid	P.O. Box 23
'Plantyn'	1975	Complex hybrid	6700 AA Wageningen, Netherlands
'Clusius'	1983	Complex hybrid	
'Sapporo Autumn Gold'	1975	*U. pumila* × *U. japonica*	E. B. Smalley Dept. of Plant Pathology University of Wisconsin Madison, Wis. 53706
'Regal'	1983	Complex hybrid	
'Recerta'	1984	*U. carpinifolia* × *U. pumila*	
'Urban'	1976	Complex hybrid	L. R. Schreiber and A. M. Townsend
'Homestead'	1984	Complex hybrid	USDA Nursery Crops Research Laboratory
'Pioneer'	1984	*U. glabra* × *U. carpinifolia*	359 Main Road Delaware, Ohio 43015
'Jacan'	1979	*U. japonica*	W. G. Ronald Morden Research Station Morden, Manitoba ROG 1JO Canada
'Dynasty'	1984	*U. parvifolia*	F. S. Santamour U.S. National Arboretum Washington, D.C. 20002
'Thomson'	1979	*U. japonica*	J. A. G. Howe R.F.P.A. Tree Nursery Indian Head, Saskatoon Canada
'American Liberty'	1983	*U. americana*	J. P. Hansel Elm Research Institute Harrisville, N.H. 03450

removed.[104] However, use of this therapy is limited due to its high cost. Ergosterol biosynthesis inhibitors, such as fenpropimorph, were recently shown to inhibit the yeast-mycelium conversion of *O. ulmi,* and to force the fungus into the yeast phase at concentrations as low as 0.03 mg/ℓ. This may be of practical importance as only hyphae are known to penetrate pit membranes (compare Figure 1 and References 25, 58, 77). Most probably *O. ulmi* is not able to colonize the xylem when forced into its yeast phase. Preliminary field trials in the Netherlands showed that fenpropimorph and some of its salts indeed arrested development of Dutch elm disease symptoms completely.[48] Due to the low concentrations necessary for inhibition of the pathogen, only relatively small amounts need to be injected. This would potentially make this treatment rather cheap in comparison with thiabendazoles. Alternative ways to administer the chemical, such as insertion of "pills" that slowly release the active ingredient into the wood, are being tested.

If an antagonist could be established into the xylem of the tree, it would be in the same environment as *O. ulmi* which, as pointed out earlier, is actually confined to the xylem vessels. In such a situation antagonism, or competition for nutrients hampering development and spread of a vascular pathogen, is easily conceivable. And, once established, the antagonist might be effective for a number of years, thus hopefully keeping the cost of such a biological control at a minimum.

Fungal control of *O. ulmi* has been attempted by placing pellets of the fungus *Trichoderma viride* into the trunks of elms.[70] In the same way, a combination of the fungi *T. viride* and *Scytalidium sp.*, the latter which is known to produce the antifungal compound scytalidin,[86] has been used.[103,105] Although *Trichoderma* species have proven to be useful for achieving biological control in other systems,[65] evidence for Dutch elm disease control is so far conflicting. Data suggest a preventive effect rather than a curative one, whilst the disease rate (i.e., the number of newly infected trees per year) must be relatively low (5% or less?).

The pathogen might also be used to control itself. Inoculation of elms showing moderate resistance with a combination of an aggressive and a nonaggressive strain of *O. ulmi* resulted in fewer disease symptoms than inoculation with the aggressive strain only, presumably because of crossprotection or induced resistance in the host.[78] Hubbes and Jeng[44] found a similar effect in American elm after inoculation with a nonaggressive strain followed 4 weeks later by a challenge inoculation with an aggressive strain.

Two other possibilities should also be mentioned. Brasier[10,11] found a so-called d-factor which affects the fitness of *O. ulmi*. Webber and Brasier[95] suggested that incompatibility between *O. ulmi* strains might be exploited. Both the d-factor and vegetative incompatibitity might slow down the overall epidemic, but no experimental evidence is yet available.

Bacterial antagonism in vitro against *O. ulmi* was observed by Hendrickx and Ledoux as early as 1937. Later, other examples of antagonism against *O. ulmi* in vitro were presented by Pomerleau and Lechevalier,[67] Szkolnik,[88] Lechevalier et al.,[53] Holmes,[41] Jewell,[47] Semer,[80] and Mazzone et al.[56] However, suppression of *O. ulmi* within the living tree was originally demonstrated by Myers and Strobel.[62] They initially selected *Pseudomonas syringae* as a potential antagonist because it can colonize internal tissues of wood, leaves, and bark.[17] *P. syringae* produces various antibiotics including syringomycin and syringotoxin,[82] which act as wide spectrum biocides against fungi and bacteria.[18] Although it is a pathogen with a wide host range,[85] it is neither a reported pathogen of elm, nor a vascular pathogen. On the basis of antimycotic activity in vitro, two saprophytic isolates were selected, one from barley and one from pear. Washed bacterial cells were placed by a gravity flow technique into one-year-old American elm seedlings in the greenhouse. A challenge inoculation with *O. ulmi* resulted in a significant reduction in the amount of vascular discoloration due to the pathogen in the elms treated with the bacteria as compared to the elms inoculated with only *O. ulmi*.

In contrast with the limited possibilities in the U.S. to conduct field studies in which elms are actually infected with *O. ulmi*, several extensive research programs in the Netherlands were conducted. Both curative and preventive treatments were compared in several elm clones, together with different techniques used to administer the various bacterial isolates.[73-75]

In Commelin elms, preventive bacterial treatments followed by inoculation with an aggressive isolate of *O. ulmi* resulted in spectacular differences in Dutch elm disease symptom development compared with control trees that were inoculated only with the pathogen. However, the method of application of the bacteria was crucial. After pressure injection, according to standard fungicide injection procedures, no decrease in symptom development was observed. In contrast, a gouge procedure by which small volumes of bacteria were

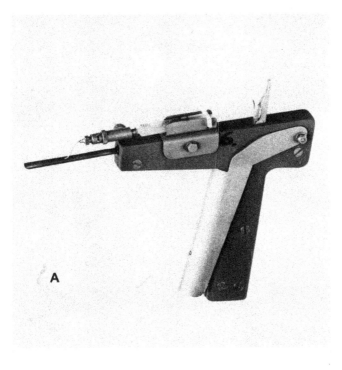

A

FIGURE 2. (A) The gouge gun as developed for treating elms with bacteria. A disposable syringe is clipped onto the main body of the device. A valve allows for automatic filling of the syringe from a bacterial supply. Normally, bacteria are shipped in 1 ℓ mixing bags (standard hospital supplies). (B) This shows the system as used in the field. The bag is connected to the gouge gun with a transfusion system with a tube length of 1.5 m.

introduced into the newest annual ring resulted in an average of 70% of trees not showing any Dutch elm disease symptoms by the end of the second season following treatment and inoculation, compared with 9% in the control groups. By then, 70% of the control trees died, compared with an average of 10% of the preventively gouge-treated ones. This procedure, for which a "gouge gun" was developed, allows one person to perform the treatments (Figure 2). The gouge technique was used for a series of other experiments, including ones in which only natural infection was allowed to occur over the course of several years. These latter experiments were, ironically, disturbed by the efficiency of the sanitation program in the Netherlands. For instance, in one such trial comprising approximately 14,000 elms at many different locations in the Netherlands (Figure 3), the overall losses due to Dutch elm disease in the control groups were only 0.68% during the second season after the bacterial treatments were performed. This figure was unexpectedly low, even for those enforcing the expensive sanitation program. In the comparable groups of trees treated with bacteria the losses were even lower, 0.26% or 0.48%, depending on the bacterial isolate used (preliminary data). In a similar 7,000 tree experiment, in Amsterdam, losses due to Dutch elm disease were 1.1% in the controls and 0.3% in the treated trees during 1984.

The longevity of the protective effect is not yet known, but we assume it to be related to the population of the antagonistic bacteria in the current growth ring, because in the ring-porous wood of elm most of the water transport takes part in the current ring. This immediately

FIGURE 2B.

lead to the question of the identity of bacteria applied to a tree and the ones recovered from it. Recently, Lam et al.[51] showed that the recombinant plasmid pRK2013 constructed by Figurski and Helinski[27] can be used as a suicide plasmid that introduces its kanamycin-resistant transposon Tn903 (which is identical to Tn601 and Tn55[64]) into a number of plant-associated pseudomonads. Southern hybridization studies showed that only a small portion of the plasmid, coinciding with the location of the transposon, was present in the kanamycin-resistant *Pseudomonas* derivatives. The plasmid sequences appeared to be inserted at a number of different sites in the recipient genome (Figure 4A). Presently, studies are underway in which strains of *Pseudomonas* marked with the transposon, or alternatively, with a plasmid-based metabolic marker[20] were introduced into elms. Transposon-bearing pseudomonads could be recovered from American elms in the greenhouse 2 years after the introduction of a transposon-bearing strain, showing the feasibility of this approach to demonstrating the identity of the bacteria introduced into the tree and the ones recovered from it (Figure 4B). More data on bacterial populations in elm are needed, together with data from field experiments such as the one mentioned in Figure 3, to allow for assessment of the longevity of the protective effect.

The other main unsolved question is what is the basis of the protective effect. Some speculation has been made that the suppression of vascular discoloration, resulting from *O.*

FIGURE 3. Locations of elms within a country-wide (Netherlands) field experiment on biological control of Dutch elm disease. Approximately 14,000 trees at 62 locations were included; half of them were treated with bacteria and natural infections were monitored.

ulmi infections in elms by pseudomonads, was due to antimycotic(s) produced by them. Myers and Strobel[62] noted that two nonantimycotic-producing mutants of *P. syringae* were ineffective in causing a significant reduction in vascular discoloration by *O. ulmi*. Whether these antimycotic(s) inhibit *O. ulmi* metabolically or by effectively chelating ions, such as Fe^{3+}, is still under investigation.

Many different pseudomonads induce similar protective effects in elms against *O. ulmi*, although they clearly differ in the antibiotics they produce. Furthermore, *O. ulmi* usually can be recovered from trees in the preventive trials and the protective effect is usually observed in elms showing some resistance. However, this preventative effect was not observed in the very susceptible field elm (*Ulmus carpinifolia*). This raises the question about the prospect that something other than antibiotics and/or competition for specific compounds is involved in conferring the protective effect in elm.

In our view the most attractive alternative hypothesis would be that the bacterial treatments actually induce or enhance the resistance of the tree. The phenomenon of induced resistance was recognized before the discovery of Dutch elm disease[2] and has intrigued plant pathologists ever since. In 1933, Chester postulated that "In ornamental plantings ... a serious consideration of the application of the findings of acquired immunity (as he called the

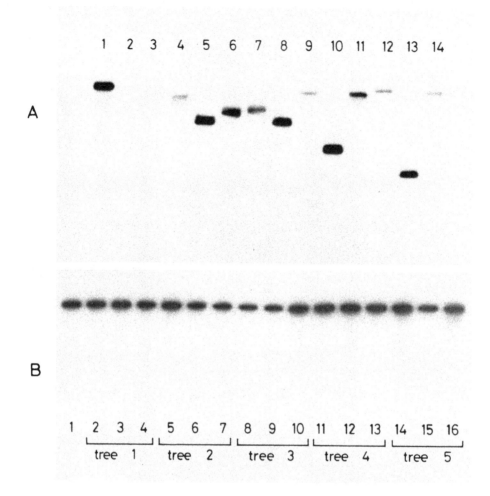

FIGURE 4. Autoradiograms of Southern blots from kanamycin-resistant *P. syringae* derivatives marked with a small portion of pRK2013, coinciding with the location of the transposon Tn*903*. The probe used was the ca. 5,000 basepair fragment containing the 3094 basepair transposon. (A) Lane 1, the donor parent HB101 (pRK2013); lane 2, HB101; lane 3, the recipient parent MSU174; lanes 4-14, MSU174 kanamycin-resistant derivatives. Insertion of the plasmid fragment apparently was at a number of different sites in the *P. syringae* DNA (lanes 4-14), but no homology was observed with the parent or recipient genome (lanes 2 and 3). (From Lam, S. T., Lam, B. S., and Strobel, G., *Plasmid*, 13, 200, 1985. With permission.) (B) Lane 1, one of the kanamycin-resistant MSU174 derivatives; lane 2-16, kanamycin-resistant reisolations (after 2 growing seasons) from five elms inoculated with the derivative shown in lane 1. The apparently stable insert identifies these reisolations (Lam, S. T., et al., unpublished).

phenomenon) is warranted''. He added that future research should focus, amongst other points, on ''the nature of acquired immunity in plants''.[16] Since then, much work was done on this topic,[49,50] but the nature of induced resistance is still not well understood. However, the occurrence of induced resistance is nowadays widely accepted and even exploited.[35]

We have already mentioned that induction of resistance in elm was made plausible using nonaggressive and aggressive strains of *O. ulmi*. Scheffer et al.[78] observed an effect only in an elm clone of moderate resistance (*U. × hollandica* clone 390), but not in the susceptible Belgian elm (*U. × hollandica* 'Belgica'). Hubbes and Jeng,[44] on the other hand, were able to induce a resistance response in American elm, which is commonly considered to be extremely susceptible. It may be, however, that a certain degree of resistance can be induced in the American elm, but probably less so in the susceptible Belgian elm. Some caution is necessary here though, as the different experiments cannot be strictly compared.

In the experiments on biological control with *Trichoderma*, which is not known to produce any antibiotics, protective effects were observed after preventive treatments. Pseudomonads also produced protective effects in the Dutch field experiments. Curative treatments were tested, but proved unsuccessful. Primarily, in these Dutch experiments, elms were used that show a moderate resistance, such as the Commelin elm. In an experiment with the very susceptible field elm no protective effect was observed whatsoever. Could this mean that the field elm is too susceptible, that no sufficient resistance could be induced by the bacterial treatments? And also, that once a tree is infected by *O. ulmi,* the induction of resistance is not strong enough an event to actually combat the disease?

The protection shown in American elm seedlings by Myers and Strobel[62] and in mature American elms by Murdoch et al.[60,61] would fit into this hypothesis, and even be predictable because, as pointed out, induction of resistance was shown in this elm species.

VII. CONCLUSIONS

Obviously, many questions remain about the biological control of Dutch elm disease. However, some conclusions can be reached from the foregoing discussion. There are a wide series of controls available, based on suppression of the vector population, hampering spread or suppression of the fungus, replacing elms with resistant ones, or combinations of the above, often called "integrated control". However, generally the cost of these control measures is such that many elm owners, governmental, institutional, and private, often have to accept losing their elms.

In the final analysis, elms with a high degree of resistance will hopefully replace many trees lost to Dutch elm disease. To save many of the already existing elms a relatively cheap control should be available. Hopefully, the biological control of the pathogen may become such an alternative.

REFERENCES

1. **Beattie, R. K.,** How the Dutch elm disease reached America, *Proc. Natl. Shade Tree Conf.,* 9, 101, 1933.
2. **Beauverie, J.,** Essais d'immunization des végétaux contre les maladies cryptogamiques, *C. R. Acad. Sci. (Paris),* 133, 107, 1901.
3. **Beckman, C. H.,** Production of pectinase, cellulase and growth promoting substance by *Ceratocystis ulmi, Phytopathology,* 46, 605, 1956.
4. **Benhamou, N., LaFontaine, J. G., Joly, J. R., and Ouellette, G. B.,** Ultrastructural localization in host tissues of a toxic glycopeptide produced by *Ophiostoma ulmi,* using monoclonal antibodies, *Can. J. Bot.,* 63, 1185, 1985.
5. **Benhamou, N., Ouellette, G. B., LaFontaine, J. G., and Joly, J. R.,** Use of monoclonal antibodies to detect a phytotoxic glycopeptide produced by *Ophiostoma ulmi,* the Dutch elm disease pathogen, Can. J. Bot., 63, 1177, 1985.
6. **Betrem, J. C.,** De iepenziekte en de iepenspintkever, *Tijdschr. Plantenziekten,* 35, 273, 1929.
7. **Biehn, W. L. and Dimond, A. E.,** Effect of pectin source and sugars on polygalacturonase production by *Ceratocystis ulmi, Phytopathology,* 61, 745, 1971.
8. **Bonsen, K. J. M., Scheffer, R. J., and Elgersma, D. M.,** Barrier zone formation as a resistance mechanism of elms to Dutch elm disease, *IAWA Bull. New Ser.,* 6, 71, 1985.
9. **Brasier, C. M.,** Genetics of pathogenicity in *Ceratocystis ulmi* and its significance for elm breeding, in *Resistance to Diseases and Pests in Forest Trees,* Heybroek, H. M., Stephan, B. R., and von Weissenberg, K., Eds., Pudoc, Wageningen, the Netherlands, 1982, 224.

10. **Brasier, C. M.,** The future of Dutch elm disease in Europe, in *Research on Dutch Elm Disease in Europe,* Burdekin, D. A., Ed., Forestry Commission Bulletin No. 60, Her Majesty's Stationery Office, London, 1983, 96.

11. **Brasier, C. M.,** A cytoplasmatically transmitted disease of *Ceratocystis ulmi, Nature (London),* 305, 220, 1983.

12. **Buisman, C. J.,** The Dutch elm disease, *Phytopathology,* 20, 111, 1930.

13. **Buisman, C. J.,** *Ceratostomella ulmi,* de geslachtelijke vorm van *Graphium ulmi* Schwarz, *Tijdschr. Plantenziekten,* 38, 1, 1932.

14. **Campana, R. J.,** Eradicative pruning, in *Dutch Elm Disease, Perspectives After 60 Years,* Sinclair, W. A., and Campana, R. J., Ed., Cornell University Agricultural Experiment Station, Search (Agriculture), 8, 1978, 33.

15. **Chapman, J. W.,** The introduction of a European Scolytid, the smaller elm bark beetle, *Scolytus multistriatus* (Marsh.) into Massachusetts, *Psyche,* 17, 63, 1910.

16. **Chester, K. S.,** The problem of acquired physiological immunity in plants, *Q. Rev. Biol.,* 8, 129, 1933.

17. **Crosse, J. E.,** Epidemiological relations of the Pseudomonad pathogens of deciduous fruit trees, *Annu. Rev. Phytopathol.,* 4, 291, 1966.

18. **DeVay, J. E., Gonzales, C. F., and Wakeman, R. J.,** Comparison of the biocidal activities of syringomycin and syringotoxin and the characterization of isolates of *Pseudomonas syringae* from citrus hosts, Proc. 4th Int. Conf. Plant Pathogenic Bacteria, Angers, France, 1978, 643.

19. **Dimond, A. E.,** The origin of symptoms of vascular wilt diseases, in *Phytotoxins in Plant Diseases,* Wood, R. K. S., Ballio, A., and Graniti, A., Eds., Academic Press, London, 1972, 289.

20. **Drahos, D., Hemming, B., McPherson, S., and Brackin, J.,** β-Galactosidase, a selectable non-antibiotic chromogenic marker for fluorescent pseudomonads, *Phytopathology,* 74 (Abstr.), 800, 1984.

21. **Elgersma, D. M.,** Resistance mechanisms of elms to *Ceratocystis ulmi, Meded. Phytopathol. Lab. "Willie Commelin Scholten",* 77, 1, 1969.

22. **Elgersma, D. M.,** Production of pectic and cellulolytic enzymes by aggressive and non-aggressive strains of *Ophiostoma ulmi, Neth. J. Plant Pathol.,* 82, 161, 1976.

23. **Elgersma, D. M.,** Resistance mechanisms of elms to Dutch elm disease, in *Resistance to Diseases and Pests in Forest Trees,* Heybroek, H. M., Stephan, B. R., and von Weissenberg, K., Eds., Pudoc, Wageningen, the Netherlands, 1982, 143.

24. **Elgersma, D. M.,** Host-parasite interactions in Dutch elm disease, in *Research on Dutch Elm Disease in Europe.* Proc. European Economic Community Research Seminar, Guernsey, Channel Islands, Burdekin, D. A., Ed., Forestry Commission Bulletin No. 60., Her Majesty's Stationery Office, London, 1983, 78.

25. **Elgersma, D. M., and Steerenberg, P. A.,** A study of the development of *Ophiostoma ulmi* in elms with immunological techniques, *Neth. J. Plant Pathol.,* 84, 127, 1978.

26. **Epstein, A. H.,** Preventing root-graft transmission, in *Dutch Elm Disease, Perspectives After 60 Years,* Sinclair, W. A. and Campana, R. J., Eds., Cornell University Agricultural Experiment Station, Search (Agriculture), 8, 1978, 32.

27. **Figurski, D. H., and Helinski, D. R.,** Replication of an origin-containing derivative of plasmid RK2 dependent on a plasmid function provided in trans, *Proc. Natl. Acad. Sci. USA,* 76, 1648, 1979.

28. **Fransen, J. J.,** Enkele gegevens omtrent de verspreiding van de door *Graphium ulmi* Schwarz veroorzaakte iepenziekte door de iepenspintkever, *Eccoptogaster (Scolytus) scolytus* F. en *Eccoptogaster (Scolytus) multistriatus* Marsh. in verband met de bestrijding dezer ziekte, *Tijdschr. Plantenziekten.* 37, 49, 1931.

29. **Fransen, J. J.,** De kleine iepenspintkever *Scolytus (Eccoptogaster) multistriatus* Marsh. als verbreider der iepenziekte, *Tijdschr. Plantenziekten,* 38, 197, 1932.

30. **Fransen, J. J.,** Iepenziekte, iepenspintkevers en beider bestrijding, Comité inzake de bestudeering en bestrijding van de iepenziekte, mededeeling No. 32, 1939.

31. **Fransen, J. J. and Buisman, C. J.,** Infectieproeven op verschillende iepensoorten met behulp van iepenspintkevers, *Tijdschr. Plantenziekten,* 41, 221, 1935.

32. **Gkinis, A. and Stennes, M.,** How to inject elms with systemic fungicides, Extension Folder 504, Agricultural Extension Service, University of Minnesota, St. Paul, Minn., 1980.

33. **Guyot, M.,** Notes de pathologie végétale, *Bull. Soc. Pathol. Végétale Fr.,* 8, 132, 1921.

34. **Hendrickx, F. L. and Ledoux, P.,** Sur l'antagonisme existant entre une bactérie et l'agent (*Ophiostoma ulmi* (Schwarz) Nannfeldt) de la maladie de l'orme (*Ulmus sp.*), *C. R. Mem. Soc. Biol.,* 126, 99, 1937.

35. **Hepburn, A. G., Wade, M., and Fraser, R. S. S.,** Present and future prospects for exploitation of resistance in crop protection by novel means, in *Mechanisms of Resistance to Plant Diseases,* Fraser, R. S. S., Ed., Martinus Nijhoff/Dr. W. Junk Publishers, Dordrecht, the Netherlands, 1985, 425.

36. **Heybroek, H. M.,** Elm breeding in the Netherlands, *Silvae Genet.,* 6, 112, 1957.

37. **Heybroek, H. M.,** The future of the elm in North America, in *Proc. 43rd Annu. Michigan Forestry Park Conf.,* Michigan State University, East Lansing, Mich., 1969, 54.

38. **Heybroek, H. M.,** Three aspects of breeding trees for disease resistance, in *Proc. 2nd World Consultation on Forest Tree Breeding,* Food and Agriculture Organization, Rome, 1970, 519.

39. **Heybroek, H. M., Elgersma, D. M., and Scheffer, R. J.,** De iepeziekte: een ecologisch ongeluk, *Nat. Tech.,* 49, 604, 1981.

40. **Heybroek, H. M., Elgersma, D. M., and Scheffer, R. J.,** Dutch elm disease: an ecological accident, *Outlook Agric.,* 11, 1, 1982.

41. **Holmes, F. W.,** The Dutch elm disease as investigated by the use of tissue cultures, antibiotics, and pectic enzymes, Ph.D. thesis, Cornell University, Ithaca, N.Y., 1954.

42. **Holmes, F. W., Demaradski, J. S., Clark, H. S., Cox, A. P., Feldman, W. C., Jr., and Kuzmiski, F. T.,** Pectic enzymes and Dutch elm disease, *Bull. Mass. Agric. Exp. Stn.,* 518, 33, 1959.

43. **Hoog, G. S. de and Scheffer, R. J.,** *Ceratocystis* vs. *Ophiostoma:* a reappraisal, *Mycologia,* 76, 292, 1984.

44. **Hubbes, M. and Jeng, R. S.,** Aggressiveness of *Ceratocystis ulmi* strains and the induction of resistance in *Ulmus americana, Eur. J. For. Pathol.,* 11, 257, 1981.

45. **Hunt, J.,** Taxonomy of the genus *Ceratocystis, Lloydia,* 19, 1, 1956.

46. **Husain, A. and Dimond, A. E.,** The function of extracellular enzymes of Dutch elm disease pathogen, *Proc. Natl. Acad. Sci. USA,* 44, 594, 1958.

47. **Jewell, T. R.,** Investigations on microbial antagonism to *Ceratocystis ulmi* (Buism.) C. Moreau *in vitro,* M.S. thesis, University of Maine, Orono, Maine, 1967.

48. **Kerkenaar, A., Scheffer, R. J., Nielander, H. B., Brakenhoff, A. C., and Elgersma, D. M.,** On the chemical control of *Ophiostoma ulmi* with sterol biosynthesis inhibitors *in vitro* and *in vivo,* Abstr. 6th Int. Cong. Pesticide Chemistry, IUPAC, Ottawa, Canada, August 10 to 15, 1986, 3C-03.

49. **Kuć, J.,** Plant-immunization mechanisms and practical implications, in *Active Defense Mechanisms in Plants,* Wood, R. K. S., Ed., Plenum Press, New York, 1982.

50. **Kuć, J.,** Induced immunity to plant disease, *BioScience,* 32, 854, 1982.

51. **Lam, S. T., Lam, B. S., and Strobel, G. A.,** A vehicle for the introduction of transposons into plant associated pseudomonads, *Plasmid,* 13, 200, 1985.

52. **Lanier, G. N. and Epstein, A. H.,** Vector suppression, in *Dutch Elm Disease, Perspectives After 60 Years,* Sinclair, W. A., and Campana, R. J., Eds., Cornell University Agricultural Experiment Station, Search (Agriculture), 8, 1978, 30.

53. **Lechevalier, H., Acker, R. F., Corke, C. T., Haenseler, C. M., and Waksman, S. A.,** Candicidin, a new antifungal antibiotic, *Mycologia,* 45, 155, 1953.

54. **Lester, D. I. and Smalley, E. B.,** Response of *Ulmus pumila* and *U. pumila* × *rubra* hybrids to inoculation with *Ceratocystis ulmi, Phytopathology,* 62, 848, 1972.

55. **May, C.,** The Dutch elm disease, *Proc. Nat. Shade Tree Conf.,* 6, 91, 1930.

56. **Mazzone, H. M., Kluck, J., Dubois, N. R., and Zerillo, R.,** Dutch elm disease control with biological agents or their metabolites, in *Proc. Dutch Elm Disease Symp. Workshop,* Kondo, E. S., Hiratsuka, R., and Denyer, W. B. G., Eds., Manitoba Department of Natural Resources, Winnipeg, Manitoba, 1982, 36.

57. **Melin, E. and Nannfeldt, J. A.,** Researches into the blueing of ground woodpulp, *Sven. Skogsvardforen. Tidskr.,* 32, 397, 1934.

58. **Miller, H. J. and Elgersma, D. M.,** The growth of aggressive and nonaggressive strains of *Ophiostoma ulmi* in susceptible and resistant elms, a scanning electron microscopial study, *Neth. J. Plant Pathol.,* 82, 51, 1976.

59. **Moreau, C.,** Coexistence des formes *Thielaviopsis* et *Graphium* chez une souche de *Ceratocystis major* (van Beyma) nov. comb. Remarques sur les variations des *Ceratocystis, Rev. Mycol.,* 17 (Suppl. Colonial 1), 17, 1952.

60. **Murdoch, C. W., Campana, R. J., and Hoch, J.,** On the biological control of *Ceratocystis ulmi* with *Pseudomonas fluorescens, Phytopathology,* 74 (Abstr.), 805, 1984.

61. **Murdoch, C. W., Campana, R. J., and Hoch, J.,** Development of Dutch elm disease inhibited by fluorescent pseudomonads, *Biol. Cult. Tests,* 1, 71, 1986.

62. **Myers, D. F. and Strobel, G. A.,** *Pseudomonas syringae* as a microbial antagonist against *Ceratocystis ulmi* in the apoplast of American elm, *Trans. Br. Mycol. Soc.,* 80, 389, 1983.

63. **Neely, D. and Himelick, B.,** Effectiveness of Vapam in preventing root graft transmission of the Dutch elm disease fungus, *Plant Dis. Rep.* 49, 106, 1965.

64. **Oka, A., Sugisaki, H., and Takanami, M.,** Nucleotide sequence of the kanamycin resistance transposon Tn *903, J. Mol. Biol.,* 147, 217, 1981.

65. **Papavizas, G. C.,** *Trichoderma* and *Gliocladium:* biology, ecology, and potential for biocontrol, *Annu. Rev. Phytopathol.,* 23, 23, 1985.

66. **Pomerleau, R.,** Pathological anatomy of the Dutch elm disease. Distribution and development of *Ceratocystis ulmi* in the elm tissues, *Can. J. Bot.,* 48, 2043, 1970.

67. **Pomerleau, R. and Lechevalier, H.,** Etude de l'effect antibiotique d'une bactérie sur le developpement du *Ceratostomella ulmi* (Schwarz) Buisman, *Rev. Can. Biol.,* 6, 478, 1947.
68. **Rackham, O.,** *Ancient Woodland,* Edward Arnold, London, 1980, 255.
69. **Rebel, H.,** Phytotoxins of *Ceratocystis ulmi,* Thesis, State University Utrecht, the Netherlands, 1969.
70. **Ricard, J. L.,** Field observations on the biocontrol of Dutch elm disease with *Trichoderma viride* pellets, *Eur. J. Forest Pathol.,* 13, 60, 1983.
71. **Roepke, W.,** Verdere gegevens omtrent de iepenziekte en de iepenspintkever, *Tijdschr. Plantenziekten,* 36, 231, 1930.
72. **Scheffer, R. J.,** Toxins in Dutch elm disease, in *Research on Dutch Elm Disease in Europe,* Proc. Eur. Econ. Community Research Seminar, Burdekin, D. A., Ed., Forestry Comm. Bull. No. 60., Her Majesty's Stationery Office, London, 1983, 82.
73. **Scheffer, R. J.,** Biological control of Dutch elm disease by *Pseudomonas species, Ann. Appl. Biol.,* 103, 21, 1983.
74. **Scheffer, R. J.,** *Pseudomonas* treatments as a possible control method for Dutch elm disease, in *Proc. 24e Colloqué de la Société Française de Phytopathologie,* Les Colloqués de l'INRA no. 18, Bordeaux, May 26 to 28, 1983.
75. **Scheffer, R. J.,** Dutch elm disease. Aspects of pathogenesis and control, Thesis, University of Amsterdam, the Netherlands, 1984.
76. **Scheffer, R. J. and Elgersma, D. M.,** Detection of a phytotoxic glycopeptide produced by *Ophiostoma ulmi* in elm by enzyme-linked immunospecific assay (ELISA), *Physiol. Plant Pathol.,* 18, 27, 1981.
77. **Scheffer, R. J. and Elgersma, D. M.,** A scanning electron microscope study of cell wall degradation in elm wood by aggressive and non-aggressive isolates of *Ophiostoma ulmi, Eur. J. Forest Pathol.,* 12, 25, 1982.
78. **Scheffer, R. J., Heybroek, H. M., and Elgersma, D. M.,** Symptom expression in elms after inoculation with combination of an aggressive and a nonaggressive strain of *Ophiostoma ulmi, Neth. J. Plant Pathol.,* 86, 315, 1980.
79. **Schwarz, M. B.,** Das Zweigsterben der Ulmen, Trauerweiden und Pfirsichbäume. *Meded. Phytopathol. Lab.* "Willie Commelin Scholten", 5, 1, 1922.
80. **Semer, C. R.,** Biological control of Dutch elm disease, M.S. thesis, Ohio State University, Columbus, Ohio, 1978.
81. **Sinclair, W. A.,** Range, suscepts, losses, in *Dutch Elm Disease, Perspectives After 60 Years,* Sinclair, W. A., and Campana, R. J., Eds., Cornell University Agricultural Experiment Station, Search (Agriculture), 8, 1978.
82. **Sinden, S. L., DeVay, J. E., and Backman, P. A.,** Properties of syringomycin, a wide spectrum antibiotic and phytotoxin produced by *Pseudomonas syringae,* and its role in the bacterial canker disease of peach trees, *Physiol. Plant Pathol.,* 1, 199, 1971.
83. **Spierenburg, D.,** Een onbekende ziekte in de iepen. I, *Versl. Meded. Plantenziektenkundige Dienst Wageningen,* 18, 3, 1921.
84. **Spierenburg, D.,** Een onbekende ziekte in de iepen. II, *Versl. Meded. Plantenziektenkundige Dienst Wageningen,* 24, 3, 1922.
85. **Stapp, C.,** *Bacterial Plant Pathogens,* Oxford University Press, London, 1961.
86. **Stillwell, M. A., Wall, R. E., and Strunz, G. M.,** Production, isolation and antifungal activity of scytalidin, a metabolite of *Scytalidium sp., Can. J. Microbiol.,* 19, 597, 1973.
87. **Svaldi, R. and Elgersma, D. M.,** Further studies on the activity of cell wall degrading enzymes of aggressive and non-aggressive isolates of *Ophiostoma ulmi, Eur. J. Forest Pathol.,* 12, 29, 1982.
88. **Szkolnik, M.,** Antagonistic activity of a species of *Actinomyces* against *Ceratostomella ulmi in vitro, Phytopathology,* 38, 85, 1948.
89. **Takai, S., Richards, W. C., and Stevenson, K. J.,** Evidence for the involvement of cerato-ulmin, the *Ceratocystis ulmi* toxin, in the development of Dutch elm disease, *Physiol. Plant Pathol.,* 23, 275, 1983.
90. **Van Alfen, N. K. and MacHardy, W. E.,** Symptoms and host-pathogen interactions, in *Dutch Elm Disease, Perspectives After 60 Years,* Sinclair, W. A. and Campana, R. J., Eds., Cornell University Agricultural Experiment Station, Search (Agriculture), 8, 1978, 20.
91. **Upadhyay, H. P.,** *A Monograph of Ceratocystis and Ceratocystiopsis,* University of Georgia Press, Athens, Georgia, 1981.
92. **VanderPlank, J. E.,** *Disease Resistance in Plants,* Academic Press, New York, 1968.
93. **VanderPlank, J. E.,** *Disease Resistance in Plants,* 2nd ed., Academic Press, New York, 1984.
94. **Webber, J.,** A natural biological control of Dutch elm disease, *Nature (London),* 292, 449, 1981.
95. **Webber, J. F. and Brasier, C. M.,** The transmission of Dutch elm disease: a study of the processes involved, in *Invertebrate-Microbial Interactions,* Anderson, J. M., Rayner, A. D. M., and Walton, D. W. H., Cambridge University Press, London, 1984, 271.
96. **Westerdijk, J.,** Is de iepenziekte een infectieziekte?, *Tijdschr. Ned. Heidemaatsch.,* 40, 333, 1928.

97. **Weijman, A. C. M. and de Hoog, G. S.,** On the subdivision of the genus *Ceratocystis, Antonie van Leeuwenhoek,* 41, 353, 1975.
98. **Wilkinson, G.,** *Epitaph for the Elm,* Arrow Books, London, 1978.
99. **Wilson, M. and Wilson, M. J. F.,** The occurrence of the Dutch elm disease in England, *Gard. Chron.,* 83, 31, 1928.
100. **Wollenweber, H. W.,** Das Ulmensterben und sein Erreger, *Graphium ulmi* Schwarz, *Nachrichtenbl. Dtsch. Pflanzenschutzdienst,* 7, 97, 1927.
101. **Wollenweber, H. W.,** Das Ulmensterben und sein Erreger. Z. *Pilzkd.* 8, 162, 1929.
102. **Woods, A. C. and Holmes, F. W.,** Extraction of fluid from healthy and Dutch-elm-diseased elm branches using hydraulic compression, *Phytopathology,* 64, 1265, 1974.
103. **Zimmermann, G.,** Versuche zur biologischen Bekämpfung der Holländischen Ulmenkrankheit mit Trichoderma-Pellets, *Nachrichtenbl. Dtsch. Pflanzenschutzdienstes (Braunschweig),* 37, 113, 1985.
104. **Stennes, M. and French, D.,** personal communication.
105. **Fairhurst, C.,** personal communication.

Chapter 9

HYPOVIRULENCE: A NATURAL CONTROL OF CHESTNUT BLIGHT

Dennis W. Fulbright, Cynthia P. Paul, and S. Westveer Garrod

TABLE OF CONTENTS

I. INTRODUCTION

Chestnut blight, caused by the fungus *Endothia parasitica* (Murr.) And. [*Cryphonectria parasitica* (Murr.) Barr] continues to be one of the worst forest diseases in North America. Estimates place the destruction at more than 3.5 billion American chestnut [*Castanea dentata* (Marsh) Bork.] trees throughout its natural range,[39,44] ranking chestnut blight as one of the few plant diseases responsible for the near annihilation of an entire species. The disease has been incurable, uncontrollable, and unmanageable in North America. The onslaught continues to reduce the once dominant, climax tree species to an understory shrub. As the disease swept through the countryside in the first half of this century, research generated considerable public interest. Soon it was evident that this disease would conquer scientific technology, disrupt forest ecosystems, and alter the destiny of industries tied closely to the chestnut tree. More recently, unless one has personally seen diseased trees, the blight could have gone unnoticed. But for those who have steadfastly waited for a cure and the return of the tree, the fungal disease remains an abusive intruder maintaining a cyclic routine that destroys sucker shoots from the roots of the trees killed in decades past.

Within the last 10 years, however, chestnut blight has once again become of scientific interest.[63] It appears that in some locations the character of the disease is changing; natural control mechanisms have resulted in the survival of chestnut trees in Europe,[3] Michigan,[28] and other locations in North America.[43] The proposed control mechanism itself has been recognized as one of the foremost naturally occurring biological controls.[12] The recovery of trees in Italy has brought back the chestnut industry in Europe.[48] In Michigan, the recovery is equally remarkable since these are the first American chestnut trees showing large-scale recovery from chestnut blight. But as chestnut trees are long-lived organisms and the blight is ever present, it may take years to determine the final outcome of the natural control process.

In this chapter we will review the evidence that the nature of chestnut blight has changed due to biological intervention in chestnut stands in Europe and America. Since several reviews on the recovery of trees in Europe already exist,[3,18,19,33,61] we will concentrate on the situation in North America, with emphasis on Michigan. We will also include a section on how these findings may be exploited to the betterment of the American chestnut tree, and our understanding of host/parasite interactions.

II. HISTORY OF CHESTNUT BLIGHT

Endothia parasitica, an ascomycete in the family Diaporthaceae, was accidentally introduced into North America on imported oriental chestnut nursery stock at the turn of this century.[36] Although *E. parasitica* was a natural part of the fungal flora associated with oriental chestnuts (*C. mollissima* and *C. crenata*), American chestnut was found to be an extremely susceptible host.[59] First observed in New York City in 1904,[50] the blight rapidly disseminated through the natural range of the American chestnut.

Once the fungus invades and becomes established in the bark, it grows rapidly, producing an elliptical, sunken canker which kills the bark and cambium, girdles the trunk, and causes wilting and death of the distal portion of the tree (Figure 1A and 1B). Under normal conditions, little wound periderm is produced and death of a large branch can occur within one or two growing seasons. Large trees usually die within 5 years (Figure 2). The fungus does not usually invade root tissue, so sprouts from the roots grow rapidly, providing a fresh source of tissue for the pathogen. These sprouts rarely live to maturity and remain understory shrubs. Since chestnut trees require cross pollination, the likelihood of two trees living to maturity in close enough proximity to provide pollination is extremely rare. Therefore, the production of seed in the natural range is almost nonexistent.

FIGURE 1.   Normal and abnormal cankers found in blighted American chestnut stands in Michigan. (A-B) Normal cankers found in declining stands. Bark is still intact in (A) and erumpent stroma are numerous; no callus is evident. Bark has fallen from canker in (B) revealing dead cambium and wood. (C-D) Abnormal cankers found on blighted, but surviving American chestnut trees in Michigan. (C) A superficial canker where the fungus has encircled the branch but girdling and death does not occur. (D) A closing canker which produces callus tissue each year at the margins of the canker. (From Fulbright, D. W., Weidlich, W. H., Haufler, K. Z., Thomas, C. S., and Paul, C. P., *Can. J. Bot.*, 61, 3164, 1983.)

FIGURE 2.   A large American chestnut can die from blight within 5 years. (A-B) Blight-infected American chestnut tree in Michigan; (A) 1976 and (B) 1981. Arrows indicate the bole of the chestnut tree. (Photos courtesy of L. Brewer.)

In North America, the tree's natural range extended from southern Maine to Alabama, west to eastern Michigan and Ohio, and down through Missouri and Arkansas. The tree was dominant in the Appalachian mountain states including Pennsylvania, West Virginia, Kentucky, Tennessee, and North Carolina. In these states the American chestnut was in its greatest numbers and largest size growing to heights of 120 feet with basal diameters of 12 feet or greater.[44] In the natural range the shrubs or sucker shoots are still quite common. Clear cutting of vegetation releases the sucker shoots, and entire mountain sides can flourish with chestnut for several years before blight destoys them again.

Because of its value, chestnut trees were planted by pioneers throughout North America including Indiana, Illinois, western Michigan,[7] Wisconsin, Minnesota, Iowa, California, Oregon, Washington, and British Columbia, Canada.[44] Even though chestnut blight has been found in all of these areas, mature trees that have escaped infection can still be found in most of these regions.

As rapidly as the disease advanced in North America, it still took decades to move through the chestnut's natural range. It was kept out of Europe for over 30 years, probably instilling a false sense of security that the Atlantic Ocean and quarantine laws would provide the needed barrier to keep the disease out of the Mediterranean chestnut forests and orchards. Some Canadians had such hopes in 1912 when they suggested that Lake Erie would protect their chestnut forests from the ravages of the blight.[55] But moving across great bodies of water proved as easy as moving across the North American continent. By 1938, chestnut blight had been introduced into Italy and the devastation began on another continent.[3]

III. DISCOVERY OF HYPOVIRULENCE

Chestnut blight caused such complete destruction of American and European chestnut biomass that any surviving tree became a curiosity and was suspected of expressing disease resistance. For that reason, it was not surprising that Italian scientists, in 1953, began studying blight-devastated Italian chestnut (*Castanea sativa*) plantations where stump sprouts were observed to be surviving infection.[8] Later, French pathologists made the determination that these trees were not resistant, but infected with pigmentless (white) strains of *E. parasitica* that were less virulent than normally-pigmented (orange), virulent strains.[31,32] These characteristics could be cytoplasmically transmitted from strain to strain of *E. parasitica* via hyphal anastomosis (fusion). When hyphae of these newly discovered strains were introduced into the bark around the margin of a canker on a European chestnut tree, the margin callused and the infection site became swollen from wound periderm. The canker ultimately closed,[28] resulting in the appearance of abnormal cankers. The phenotype of these newly discovered strains was termed "hypovirulent".[31] Research in the U.S. demonstrated that Italian and French hypovirulent strains could transmit the cytoplasmically associated phenotype to pigmented, virulent strains of *E. parasitica* found in America. The ability to induce the closing of cankers on American chestnut was also demonstrated.[63] One of the most significant findings was the discovery of double-stranded ribonucleic acid (dsRNA) molecules in the cytoplasm of hypovirulent strains.[14] Since pigmented, virulent strains did not have dsRNA, an immediate association between the hypovirulent phenotype and the presence of dsRNA was made.

In 1977, abnormal cankers like those described by the European researchers were discovered on trees in an American chestnut stand planted by pioneering farmers in Michigan.[23] An isolate recovered from one of the cankers was pigmented but appeared debilitated in culture. Analysis of dsRNA and virulence demonstrated the hypovirulent nature of this isolate. A statewide chestnut survey demonstrated that several stands of American chestnut displayed abnormal cankers.[10,28] It was soon discovered that stands of American chestnut in Michigan could be found in three situations: (1) blight-free, (2) infected and dying of blight, and (3) infected but surviving blight. Surviving stands could be found in various stages of recovery. In some stands more trees were surviving than in others. To determine if the trees were surviving blight, long-term observation was required (Figure 3).

Two types of abnormal cankers appeared on Michigan surviving trees (Figure 1C and 1D).[28] The first type was probably, at one time, a normal canker, for death of the cambium and bark was apparent. Old mycelial fans were evident on the outer surface of the secondary xylem. At some point in canker development, tissue at the margins began producing wound periderm and secondary vascular phloem (Figure 1D). The second type of abnormal canker appeared to be superficial, that is, the fungus appeared to be restricted to the bark. Destruction of the phloem or cambium was not evident. This canker could be found encircling the trunk but girdling did not occur (Figure 1C). Hypovirulent isolates of *E. parasitica* were commonly found associated with both types of cankers. Pathogenicity tests involving host tissue from both recovering and declining stands showed that resistance was not a factor in the appearance of abnormal cankers on the trees. It was concluded that hypovirulent strains within these

stands of trees were somehow responsible for the recovery and prolonged life of American chestnut trees.[28] Therefore, more was needed to be learned about these strains.

IV. CHARACTERISTICS OF HYPOVIRULENT STRAINS

The term hypovirulence has varying definitions. Simply stated, a culture is hypovirulent when the capacity to cause disease is reduced from the expected. A working definition that is more specific for the chestnut blight system has become accepted by most researchers. Hypovirulent strains of *E. parasitica* are less virulent than normal strains due to a transmissible, cytoplasmic agent. Hypovirulent strains have been frequently characterized according to their culture morphology and dsRNA content as well as virulence status. The combination of these three factors probably best describe any individual hypovirulent strain. Presently, relatively little is known about the total genetic input of hypovirulent strains and how these phenotypic characteristics arise.

A. Culture Morphology and Virulence

By studying culture morphology and virulence in various dsRNA-free and dsRNA-containing strains, Elliston[20] has shown that hypovirulent strains show dramatic variability in morphology as well as virulence. In general, he found that nearly all dsRNA-free cultures produced similar colonies in culture and large cankers in bioassays as well as in field tests. Virulence could be measured in two ways: the size of the canker in a given time period, or the size of the canker plus the production of stromata and fruiting bodies (both sexual and asexual). Isolates with dsRNA differed in culture morphology from dsRNA-free isolates and from one another. All but one of the dsRNA-containing isolates tested had reduced virulence.

This reduction was even more pronounced when virulence was measured as a combination of canker size and sporulating capacity (sexual and asexual). Using these criteria, he found that no one particular phenotype is common to all hypovirulent strains. His data also indicated that almost all dsRNA-containing strains contain some deficiency in virulence. Sometimes these deficiencies may not be expressed until 6 months after inoculation of test trees. Perhaps the most significant part of this work was the elimination of generalized characteristics such as all European hypovirulent strains are white, hypovirulent strains do not sporulate in vivo, and hypovirulent strains are always dramatically reduced in virulence.

Isolates of hypovirulent *E. parasitica* from recovering trees in Michigan also offer a wide range of culture morphologies and virulence deficiencies.[28] Although many of the hypovirulent strains from Michigan have extremely low virulence, most of the hypovirulent strains are capable of causing moderately sized cankers with the production of asexual spores on the bark of infected trees. Spores collected from these cankers harbor dsRNA and may be involved in the dissemination of hypovirulent strains.[29] Some hypovirulent strains in our collection are virulent enough to kill young seedlings and cause large cankers on immature trees, while remaining relatively benign on older trees.[70]

B. DsRNA Content

Hypovirulent strains of *E. parasitica* were first shown to contain segments of dsRNA by Day et al.[14] and Dodds.[15] These dsRNA molecules were found to vary in size and number between strains. Most strains were infected with one to four multiple segments of dsRNA in high concentrations, and several poorly defined segments in lower concentrations. These were referred to as major bands and minor bands, respectively, when visualized after gel electrophoresis. The major bands were thought to be genomic dsRNA while the minor bands may be subgenomic dsRNA.[13]

Finding dsRNA in *E. parasitica* supports the possibility that these strains are infected with mycovirus-like agents, since the chromosome of nearly all mycoviruses is composed of multi-segmented dsRNA.[9] Most mycoviruses have not been associated with abnormal fungal physiology, and the majority of these cytoplasmic agents were discovered after random electron microscopic studies of apparently healthy cultures. A few mycoviruses have been correlated to either killer-toxin production[60] or diseases of mushrooms.[9] Although hypovirulence has been described and correlated with dsRNA in other fungal species, these reports represent preliminary or incomplete work.[11,56]

Since isolated mycoviruses are unable to reinfect their host, evidence linking some mycoviruses to specific phenotypes remains purely correlative. In the case of hypovirulence, a close correlation is based on the presence of specific dsRNA segments, specific phenotypic characteristics, concerted transfer of both dsRNA and associated phenotype,[30] and curing experiments where elimination of specific dsRNA segments is correlated with changes in phenotype (Figure 4).[24] The elimination of dsRNA in hypovirulent strains of *E. parasitica* is easily accomplished by growing cultures from isolated, single-conidia. The dsRNA is not evenly distributed to all conidia. Therefore, cultures free of dsRNA can be isolated at frequencies specific for various hypovirulent strains.[21] The elimination of dsRNA from mass-culture isolates has been accomplished by growing hypovirulent strains on agar media containing cycloheximide.[24] Not all hypovirulent isolates lost dsRNA when grown in the presence of cycloheximide. However, isolates that lose dsRNA, whether through single-conidial isolation or cycloheximide treatment, grow normally and regain normal virulence.

Recovering American chestnut groves in Michigan harbor a number of phenotypically different hypovirulent strains that display diverse dsRNA banding patterns after gel electrophoresis (Figure 5).[28] Attempts have been made to characterize hypovirulent strains by studying their dsRNA banding patterns, phenotypic expression (culture morphology and virulence), and homology of their dsRNA genomes. When the dsRNA banding patterns from six Michigan hypovirulent strains were analyzed, the number of major dsRNA segments

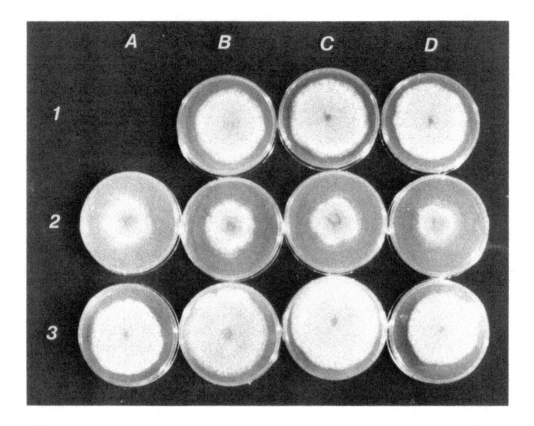

FIGURE 4. The presence of dsRNA in *Endothia parasitica* correlates with reduced virulence and abnormal culture morphology. (1B-1D) Normal virulent strains CL1, CL3, and CL4. (2A) Hypovirulent strain, GHU4, which was used to convert the normal virulent strains found in row 1 to the hypovirulent form. (2B-2D) After the dsRNA from GHU4 was transferred to the virulent strains in Row 1, they took on the characteristics of GHU4. (3A-3D) after eliminating the dsRNA with cycloheximide treatment, the normal morphology appeared and virulence was restored. (From Fulbright, D. W., *Phytopathology*, 74, 722, 1984.)

varied from two to four per strain. Many dsRNA bands that were very faint were considered subgenomic dsRNAs and not counted as major segments. The molecular weights of the dsRNA segments varied from 0.56 to 6.2 $\times$ 10^6 (Figure 5). No single dsRNA segment, based on size, was common to all strains, but most strains contained high molecular weight segments ($> 4.4 \times 10^6$). The one exception to this was the RC1 strain which was isolated from a stand of recovering trees farther east than any of the others. DsRNA banding patterns that appeared similar after gel electrophoresis were found in two hypovirulent strains from stands of recovering trees separated by about 10 mi, indicating that these locations were infected with the same dsRNA genomes.

Since a single dsRNA segment, based on size, cannot be correlated to the hypovirulent phenotype, sequence homology relationships have been studied to determine if the dsRNA segments might still be genetically related. When the dsRNA genomes from North American strains were hybridized to European dsRNA, no genetic homology was found.[47,53] Using a dot-blot technique, L'Hostis et al.[47] found that in the strains tested, all North American dsRNA, including a strain from Michigan, were genetically related. In contrast, utilizing northern blots and dsRNA from different strains than L'Hostis et al., Paul et al.[53,54] showed that there was no homology between Michigan dsRNA and dsRNA from West Virginia and other North American locations. Hypovirulent strains from Michigan, however, in spite of the variety of banding patterns, shared sequence homology. The only exception was the

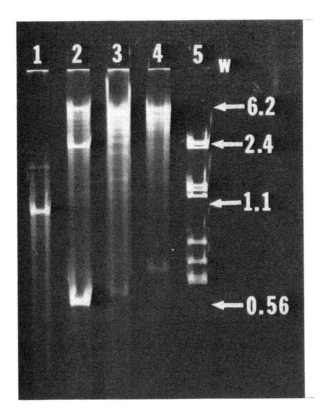

FIGURE 5. Double-stranded RNA molecules found in four representative hypovirulent strains of *Endothia parasitica* from Michigan recovering trees. (Lane 1) RC1, (Lane 2) GH2, (Lane 3) GHU4, (Lane 4) RF, and (Lane 5) REO virus. Numbers to the side represent molecular weight $\times$ 10^6. (W) Bottom of well. (From Fulbright, D. W., Weidlich, W. H., Haufler, K. Z., Thomas, C. S., and Paul, C. P., *Can. J. Bot.*, 61, 3164, 1983.)

RC1 strain which showed no homology to the dsRNAs found in hypovirulent strains from Michigan, other North American locations, or Europe.

Hypovirulent strains with several segments of dsRNA may harbor more than one virus-like agent or the various segments may just be deletion or breakdown products of one larger molecule. It has been shown that a hypovirulent strain can carry more than one cytoplasmic factor responsible for the culture morphology and virulence phenotype of the isolate. Through single-conidial isolation techniques, Elliston[21,22] demonstrated that the mass-isolate Ep60 recovered from an abnormal canker in a stand of recovering trees in western Michigan was infected with two cytoplasmic agents responsible for hypovirulence. When present singly, each agent conferred different culture abnormalities and different levels of virulence. When present together, the more debilitated agent was dominant and gave Ep60 its debilitated characteristics. Different dsRNA banding patterns on polyacrylamide gels were evident for each agent, supporting the theory that specific dsRNA segments are responsible for the associated phenotype of the mass isolate.

Transferring the dsRNA from GH2, RC1, and GHU4 (Figure 5) into one common nuclear background resulted in new cultural abnormalities and virulence deficiencies.[26] Cytoplasmic segregation was evident after single conidial isolation resulted in cultures with new morphology, reduced virulence, and new dsRNA-banding pattern profiles not seen previously in parent strains, including the appearance of new dsRNA segments.

C. Antigens

To date, no coat proteins have been associated with the dsRNA found in *E. parasitica*. The search for particles in dsRNA-infected *E. parasitica* yielded membrane-bound vesicles similar to those found in some mushroom diseases.[16] Clusters of these vesicles have been observed by electron microscopy in European hypovirulent strains.[52] Clusters of vesicles were not found in Michigan hypovirulent strains, but individual vesicles were observed.[66] Their role is not clear, but Van Alfen has attempted fusing isolated vesicles from a hypovirulent strain with fungal sphaeroplasts from a virulent strain.[62] DsRNA was apparently transferred but was very unstable, and subsequently lost upon subculture. Van Alfen's laboratory has recently isolated a replicase from European hypovirulent strains.[64] This is an important discovery since it would be the first known gene product coded for by the dsRNA genome.

D. Other Characteristics

Other characteristics of hypovirulent strains have been investigated but data are still too preliminary to be reliable. One characteristic examined was oxalate production in virulent and hypovirulent strains. Oxalic acid production is thought to be important in the virulence of some plant pathogens, and it can be found at the edge of expanding cankers on chestnut trees. Havir and Anagnostakis[36,37] found that in vitro production of oxalic acid was not detectable in hypovirulent strains, but that the same strains without the dsRNA produced detectable levels of oxalic acid.

V. TRANSMISSION

Unlike bacterial conjugative plasmids which promote their own transfer, dsRNA in fungi is dependent upon the fungal host for transfer. Since these dsRNA genomes are not known to lyse or kill fungal host cells, infection from external sources is probably remote, Therefore, dsRNA in *E. parasitica* is probably only transferred to new strains by asexual conidia or through hyphal anastomosis. In either case, the dsRNA genomes never leave the cytoplasm of the host fungus.

A. Transfer Via Spores

DsRNA has yet to be found in cultures initiated by ascospores from in vivo or laboratory crosses of hypovirulent strains. However, as mentioned above, dsRNA is commonly found in asexual pycnidiospores, commonly referred to as conidia. Conidia contain a single nucleus and cytoplasm from the mother culture which can include mycoviruses. The role these spores might play in the dissemination of hypovirulent strains will be discussed below.

B. Transfer Via Hyphal Fusion

The cytoplasmic transfer of dsRNA from one strain of *E. parasitica* to another was first proposed by Grente and Sauret.[32] However, Van Alfen et al.[63] first genetically demonstrated this mode of transfer when genetically-marked, auxotrophic strains of *E. parasitica* were paired in chestnut bark and in culture. A virulent methionine auxotroph and a hypovirulent lysine auxotroph were inoculated as pairs in the trunks of American chestnut seedlings. Methionine isolates recovered from the cankers 90 days later were hypovirulent, indicating that the hypovirulent phenotype transferred from the lysine auxotroph. More evidence for the cytoplasmic nature of hypovirulence was obtained when heterokaryons were formed after pairing hypovirulent methionine and arginine auxotrophs. Methionine and arginine auxotrophs were obtained from single-conidial isolates recovered from the heterokaryon. Both auxotrophs were hypovirulent. Therefore, cytoplasmic mixing of the two strains resulted in the hypovirulent phenotype occuring with both nuclei.

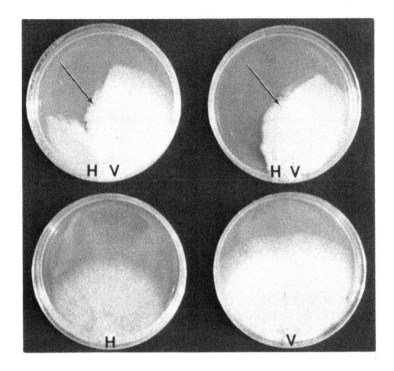

FIGURE 6.   Conversion of a virulent strain of *Endothia parasitica* with a hypovirulent strain. (H) Indicates the hypovirulent strain and (V) represents the virulent strain growing alone. (HV) Indicates the inoculum of each strain as it was initially paired on the plates. Arrows indicate areas where virulent strains, having fused with the hypovirulent strain, began to take on the characteristics of the hypovirulent strain. (From Anagnostakis, S. L. and Day, P. R., *Phytopathology*, 69, 1226, 1979.)

1. Conversion

Virulent strains of *E. parasitica* that become hypovirulent after fusion with a hypovirulent strain were said to be converted.[2] Conversion appears to be one of the most important steps in the biological control of chestnut blight. If conversion of a virulent strain occurs in a canker on a chestnut tree, the tree will often respond by producing callus tissue at the edges of the canker, forming an abnormal canker like that seen in Figure 1D. Conversions can be performed in the laboratory by pairing two strains, one hypovirulent and the other virulent, on a solidified nutrient medium (Figure 6). The virulent strain will begin to mimic the cultural and virulence phenotype of the paired hypovirulent strain. As expected, the converted strain will usually have the same dsRNA-banding pattern profile as the hypovirulent strain.[20,28]

It is still unknown if all of the abnormal cankers present in recovering groves are the result of virulent strains initiating cankers and then being converted by hypovirulent strains, or if some hypovirulent strains can initiate cankers that are ultimately controlled by tree defenses.[57]

2. Vegetative Incompatibility

Inhibiting the fusion of two strains of *E. parasitica* will prevent the conversion of virulent strains. Most filamentous fungi, including *E. parasitica*, have genetic barriers preventing the fusion of various strains of the same species.[1] Such genetic barriers are believed to be responsible for the failure of some hypovirulent strains to convert virulent strains in culture and in cankers. Vegetative incompatibility can be observed by placing two strains of *E. parasitica* very close together on solidified culture media. After a few days in the dark, the cultures are removed and left in the light. If a row of pycnidia form between the cultures,

A. Vegetative and
   Conversion Compatible

^{1}(H) ———————▶ ^{2}(V)

B. Vegetative and
   Conversion Incompatible

^{1}(H) —✖▶ ^{3}(V)

C. Vegetative Incompatible,
   but Conversion Compatible

^{2}(H) ———————▶ ^{3}(V)

D. Bridging Conversion
   Incompatibility

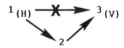

FIGURE 7.   Effect of vegetative and conversion compatibilities on the transfer of the hypovirulent phenotype between strains of *Endothia parasitica*. (H) Indicates hypovirulent strains and (V) represents virulent strains. (→) Indicates successful transfer of dsRNA while (✖) indicates the transfer was blocked.

the two strains are considered incompatible.[1] This indicates that hyphal fusion was prevented. If the cultures fuse, they are considered compatible. However, vegetative incompatibility is not as limiting as once thought. Some vegetatively incompatible strains of *E. parasitica* allow the transfer of dsRNA and the hypovirulent phenotype. Anagnostakis[4] has suggested that hypovirulence conversion compatibility is less stringent than vegetative compatibility. Whether this is the function of the dsRNA or of the hyphae and nuclear genes is not clear at this time. It has been speculated that some cytoplasmic exchange occurs even during an incompatible interaction and that this exchange may account for the transfer of the dsRNA. There are some strains, however, that will not fuse and will not transfer dsRNA to certain vegetatively incompatible strains. This can be overcome by using various strains that bridge vegetatively incompatible strains (Figure 7). Anagnostakis[4] concluded that networks of vegetative incompatibility groups exist that allow cytoplasmic exchange between various groups of incompatible strains. Research has demonstrated that more conversions take place when a mixture of hypovirulent strains representing various vegetative compatibility groups are used in conversion tests or biological control experiments.[4,42,46]

In extensive tests on the effectiveness of hypovirulent strains to convert virulent strains, Kuhlman and Bhattacharyya[45] found that 97% of 102 randomly selected virulent strains could be converted by 4 hypovirulent strains. These 102 virulent strains represented 40 vegetative compatibility groups. In another experiment, 95% of 118 virulent isolates were converted by 24 different hypovirulent strains. The 118 isolates represented 54 different vegetative compatibility groups from West Virginia, Virginia, and North Carolina. The 27 hypovirulent strains converted from 0 to 41% of the 118 virulent strains with an average conversion frequency of 15%. Therefore, in locations where a large number of vegetative compatibility groups are known to exist, such as West Virginia,[49] North Carolina,[51] and Connecticut,[6] natural conversion may be enhanced by introducing hypovirulent strains representing a significant number of the known vegetative compatibility groups in the geographic location. However, Kuhlman and Bhattacharyya[45] found that hypovirulence did not spread over a 300-meter study area even when the virulent isolates in the area were readily converted in the laboratory by the introduced hypovirulent strains. This indicated that other factors are also involved in the dissemination of hypovirulence.

VI. HYPOVIRULENCE AS AN INTRODUCED BIOLOGICAL CONTROL

As described above, normal cankers on chestnut trees can be induced to close by treatment with the proper hypovirulent strains of *E. parasitica*. This in itself is a remarkable achievement since no treatment was known previously. Grente and Berthelay-Sauret have been able to introduce hypovirulent strains into French chestnut groves and obtain measurable dissemination in 3 years.[33] This had not been possible in North America with American chestnut trees.

A. Disease Treatment

Several laboratories have experimented with the introduction of hypovirulent strains on American chestnut trees.[3,27,41,45] Treatment usually consisted of making wounds around the margin of actively expanding cankers, and filling the wounds with selected hypovirulent strains from plates of a solidified culture medium.[27] Trees with normal cankers that have been treated with hypovirulent strains have remained alive longer than control trees which were not treated with hypovirulent strains.[27,41] Many of the trees in our test plots have been treated seven or eight times since 1981 but new cankers continue to develop within these trees. Two hypovirulent strains have been continuously used in our studies: GHU4 (or its converted derivatives) and GH2. GHU4 is severly deficient in virulence and does not produce conidia on chestnut bark. Therefore, the halting of canker expansion after treatment with this strain is effective for the treated canker only. There are few ways for the dsRNA to spread from a treated canker to other cankers on the tree without the production of conidia. The GH2 strain, however, grows and sporulates in chestnut bark and produces conidia that carry dsRNA. Canker expansion is not stopped but slows considerably after canker treatment. Conidia are produced yearly at the margin of the canker. These conidia were considered to be a possible source of dsRNA for the rest of the stand (see below).

B. Establishment and Dissemination of Hypovirulent Strains

Since the discovery of the hypovirulent phenotype in *E. parasitica* and its association with dsRNA, several theories have been put forth on the origin of the cytoplasmic molecules. Perhaps they were always present in a small segment of the population and are just now being found.[17] Perhaps the dsRNA is carried in a prophage-like state within the chromosome and is induced in some environmental situations;[61] or perhaps the dsRNA has been transferred from some other fungus or microorganism and has found a new host in which to replicate.[67] In any event, it is likely that the origin of the dsRNA is several years away from being discovered. It is logical to assume that if "source" strains are found, they will be employed to help establish hypovirulent strains in declining stands of trees. Until then, experiments to elucidate the possible dissemination of hypovirulent strains in recovering stands or from introduced hypovirulent strains in declining chestnut stands need to be continued.

It is thought that the primary mode of dissemination of normal, virulent strains of *E. parasitica* is through long distance dispersal of ascospores or local spread of conidia.[38] Since dsRNA is not thought to be carried in the ascospores, this mode of spread has been considered unlikely for the dissemination of hypovirulent strains. Recently, Russin et al.[58] has demonstrated that several insects can carry propagules of the virulent and hypovirulent pathogen, indicating that insects might be involved with long distance dispersal of the fungus. Mites have also been found to carry the pathogen.[35]

Several methods have been used to study the establishment and dissemination of virulent and hypovirulent strains of *E. parasitica* in North America. Various parameters have been considered in measuring the spread of the introduced hypovirulent strains. These methods have emphasized the survival of trees, the frequency of abnormal cankers, or the ability to isolate cultures with hypovirulent-like culture morphology. In a 6-year study by Jaynes and DePalma,[41] individual tree survival was extended for treated trees when compared to nontreated trees, but overall chestnut biomass declined in treated plots at rates equal to nontreated plots. The natural spread of hypovirulent strains in this study was considered minimal.

In another study, Willey[65] inoculated healthy sprouts with 11 hypovirulent strains, representing strains from Europe and North America. Large wounds were made in the bark and then each hypovirulent strain was inoculated separately on each tree. During the next 18 months, isolates of *E. parasitica* were recovered from naturally developing secondary cankers. These were monitored for hypovirulent culture morphology, and random isolates were assayed for dsRNA and conversion capacity. Suspect hypovirulent strains were found in 19% of the 475 naturally occurring cankers. This study also demonstrated that a single canker can support the growth of both hypovirulent and virulent strains. Some hypovirulent strains were recovered that did not resemble the culture morphology of any of the 11 original hypovirulent isolates released in the study. This work, coupled with a surprising new finding that dsRNA-containing strains are already present in West Virginia chestnut stands,[17] demonstrated that studies designed to follow specific, readily identifiable hypovirulent strains within the plot needed to be pursued. Recently, dissemination and conversion experiments utilizing a brown-pigmented mutant strain carrying different dsRNAs have helped to demonstrate that vegetative compatibility can influence the rate at which an isolate or a canker is converted.[40]

In 1982, Garrod et al.[29] began to study the fate of virulent and hypovirulent strains released in a stand of declining trees in Michigan. Included in the study were three *E. parasitica* strains: CL1, a virulent strain recovered from the same grove in 1980; CL1-PCNB[R] (pentachloronitrobenzene), a naturally occurring fungicide resistant isolate; and CL1-PCNB[R] converted by the hypovirulent strain GH2. It was thought that the GH2-banding pattern profile in acrylamide gels would offer a unique fingerprint of the dsRNA, which would help in following the specific dsRNA molecules (Figure 5). The strains were grown on autoclaved chestnut segments in petri dishes with a solidified culture medium. After several days growth, copious pycnidiospores produced on the chestnut segments and culture medium were evident. The lids were removed and the dishes were tied on trees. No direct inoculations of any strains were made on the trees. Small wounds were systematically placed in the bark beneath the dishes. The plates functioned as temporary cankers and the passivly released conidia were allowed to naturally disseminate. The study plot consisted of 20 trees, with each strain placed on 5 trees including 5 control trees without inoculated dishes. The petri dishes were changed every 2 months for 6 months. *E. parisitica* isolates from cankers that developed on the trees were assayed for their dsRNA content and their sensitivity to PCNB. By the following year, dsRNA-containing cankers were recovered from 4 trees (Table 1). Of these trees, two had the hypovirulent inoculum source placed on them, while the other two trees had the virulent CL1 strain or the virulent PCNB[R] strain placed on it. In 1984, dsRNA-containing isolates were recovered from six trees. DsRNA-containing strains were recovered from two trees that did not have the hypovirulent source placed on them. In the third year of the study, 11 trees harbored dsRNA-containing strains of *E. parasitica*. Of the 11 trees, 4 had the hypovirulent inoculum source, 3 had the CL1-PCNB[R] inoculum source, and 3 had the virulent, CL1 inoculum sources originally placed on them. Of the 11 trees 1 was a control and did not have an inoculum source placed on it at any time. The dsRNA-containing strains were found in natural cankers on the trees as well as in cankers from the wound sites. A very important observation was also made in this study. The dsRNA was found not only in the PCNB[R] strain in which the dsRNA was originally placed in the stand, but also in strains sensitive to PCNB, indicating that natural conversion had occurred between strains within the stand. Although most of the dsRNA-containing isolates had the GH2 banding pattern, some carried modifications of the pattern, demonstrating that the various dsRNA segments were not always stable. The fate of these trees is now being monitored. It will be interesting to see if the dsRNA can disseminate fast enough to keep up with the secondary cankers constantly appearing on the trees. Little is known about the vegetative compatibility groups found in this stand of trees. Since this stand was blighted in the 1970s, a very

Table 1
DISSEMINATION IN A MICHIGAN
DECLINING AMERICAN CHESTNUT
STAND OF AN INTRODUCED
HYPOVIRULENT STRAIN OF
***ENDOTHIA PARASITICA* AND ITS**
dsRNA MOLECULES[29,71]

Tree	Inoculum source placed on tree[a]	1983	1984	1985
A	R	—	—	—
G	R	+	+	+
N	R	—	+	+
Q	R	—	—	—
S	R	—	—	+
D	H	+	+	+
F	H	+	+	+
I	H	—	—	—
M	H	—	+	+
T	H	—	—	+
C	S	—	—	—
E	S	+	+	+
J	S	—	—	—
K	S	—	—	+
R	S	—	—	+
B	C	—	—	—
H	C	—	—	—
L	C	—	—	+
P	C	—	—	—
U	C	—	—	—

Note: — , No hypovirulent strains isolated from cankers on tree; + Hypovirulent strain containing the dsRNA released in plot.

[a] R: virulent, PCNB resistant isolate; H: hypovirulent, PCNB resistant isolate; S: virulent, PCNB sensitive isolate; and C: control, no inoculum source placed on the tree.

simplistic vegetative compatibility system might exist which could allow for rapid movement of dsRNA introduced in the stand.[40] The introduction of a hypovirulent strain in this stand differed from the work of others in that the hypovirulent strain used in the study carried a selectable genetic marker and that the dsRNA molecules had been characterized, making them easily recognizable when assayed on gels after electrophoresis. The hypovirulent strain had the capacity to grow and sporulate in chestnut bark and had already demonstrated a dissemination capacity. The future use of these techniques should help in understanding the dissemination of dsRNA in complex ecosystems.

VII. FUTURE RESEARCH

It is difficult to draw many conclusions at this time. As hypovirulence research begins its second decade in North America, it is hoped that more emphasis will be placed on elucidating the characteristics of successfully disseminated hypovirulent strains. It will also be important to determine all the factors required for successful establishment and spread of hypovirulent isolates, and eventual recovery of American chestnut trees in stands now declining from blight. It must be remembered that these factors revolve around the fungus, the dsRNA molecules, the tree, and the environment. Studies must take into account all of these components if a complete understanding of chestnut recovery is to be achieved. It should also be emphasized that a better understanding of the normal genetic systems of *E. parasitica* is required before the hypovirulent culture will be completely understood. Recent work indicates that virulence can be altered by chromosomal genes,[5] as well as by cytoplasmic factors other than dsRNA.[25] The work reported by Double et al.[17] is also significant in that we must determine if dsRNA and hypovirulent strains can naturally disseminate without initiating recovery of the trees. What makes the system successful in some instances and fail in others?

The value of the chesnut blight research is enhanced even more when placed in the context of understanding the genetic control of virulence and plant disease. It will be important to know if dsRNA molecules produce specific products that interact with virulence factors produced by the fungus, or if the dsRNA causes general metabolic changes that are expressed as a loss of virulence. If such products are produced, are they effective in other fungal species? Can these gene products be cloned into other pathogens? It will also be of interest to determine if the dsRNA molecules can replicate in other fungal species, altering the genetic expression of virulence in other disease systems. These are just a few of the questions that need to be approached in the coming years.

If hypovirulence exists in *E. parasitica,* it probably exists in other disease systems. It may be more difficult to observe in other systems simply because there may be several reasons for host plants to survive infection. With chestnut blight, each and every surviving tree has a unique reason for survival.[34] In response to this possibility, our laboratory has begun to study hypovirulence in the stone fruit pathogen, *Cytospora* spp. A debilitated strain obtained from North Carolina[68] is deficient in virulence and appears to harbor a virus-like particle as well as dsRNA.[69] The role of the virus-like particle is now being assessed for its association with the debilitated strain and its capacity to transmit this phenotype to other *Cytospora* strains.

Since its discovery 82 years have passed, but chestnut blight remains a constant reminder of the ecological mistakes of the past and an embarrassing defeat for a technologically oriented society. Hopefully, this disease may begin to offer some atonement by providing biological insights into the nature of disease and biological control.

ACKNOWLEDGMENTS

The Michigan State University Chestnut Research Program has been supported by the Northern Nut Growers Association, the Michigan Nut Growers Association, the Michigan State University Development Fund, and by a grant from the United States Department of Agriculture, Forest Biology Competitive Grant Program (grant 85-FSTY-9-0137). We also thank Chimney Corners Resort, Frankfort, Michigan for allowing us to use their trees in many of the reported studies. Michigan Agriculture Experiment Station Journal Series Article 12462.

REFERENCES

1. **Anagnostakis, S. L.,** Vegetative incompatibility in *Endothia parasitica, Exp. Mycol.,* 1, 306, 1977.
2. **Anagnostakis, S. L. and Day, P. R.,** Hypovirulence conversion in *Endothia parasitica, Phytopathology,* 69, 1226, 1979.

3. **Anagnostakis, S. L.,** Biological control of chestnut blight, *Science,* 215, 466, 1982.

4. **Anagnostakis, S. L.,** Conversion to curative morphology in *Endothia parasitica* and restriction by vegetative compatibility, *Mycologia,* 75, 777, 1983.

5. **Anagnostakis, S. L.,** Nuclear gene mutations in *Endothia (Cryphonectria) parasitica* that affect morphology and virulence, *Phytopathology,* 74, 561, 1984.

6. **Anagnostakis, S. L. and Waggoner, P. E.,** Hypovirulence, vegetative incompatibility, and the growth of cankers of chestnut blight, *Phytopathology,* 71, 1198, 1981.

7. **Baxter, D. V. and Strong, F. C.,** Chestnut blight in Michigan, *Mich. Agric. Exp. Stn. Tech. Bull.,* 135, 18, 1931.

8. **Biraghi, A.,** Possible active resistance to *Endothia parasitica* in *Castanea sativa, in Rep. Congr. Int. Union For. Res. Org. 11th,* Rome, Italy, 1953, 643.

9. **Bozarth, R. F.,** The physico-chemical properties of mycoviruses, in *Viruses and Plasmids in Fungi,* Lemke, P. A., Ed., Marcel Dekker, New York, 1979, 43.

10. **Brewer, L. G.,** The present status and future prospect for the American chestnut in Michigan, *Mich. Bot.,* 21, 117, 1982.

11. **Castanho, B., Butler, E. E., and Shepherd, R. J.,** The association of double stranded RNA with Rhizoctonia decline, *Phytopathology,* 68, 1515, 1978.

12. **Cook, R. J. and Baker, K. F.,** *The Nature and Practice of Biological Control of Plant Pathogens,* American Phytopathological Society, St. Paul, Minn., 1983, 539.

13. **Dawson, W. O. and Dodds, J. A.,** Characterization of subgenomic double-stranded RNAs from virus-infected plants, *Biochem. Biophys. Res. Commun.,* 107, 1230, 1982.

14. **Day, P. R., Dodds, J. A., Elliston, J. E., Jaynes, R. A., and Anagnostakis, S. L.,** Double-stranded RNA in *Endothia parasitica, Phytopathology,* 67, 1393, 1977.

15. **Dodds, J. A.,** Revised estimates of the molecular weights of dsRNA segments in hypovirulent strains of *Endothia parasitica, Phytopathology,* 70, 1217, 1980.

16. **Dodds, J. A.,** Association of type 1 viral-like dsRNA with club-shaped particles in hypovirulent strains of *Endothia parasitica, Virology,* 107, 1, 1980.

17. **Double, M. L., MacDonald, W. L., and Wiley, R. L.,** Double-stranded RNA associated with the natural population of *Endothia parasitica* in West Virginia, *Phytopathology,* 75, 624, 1985.

18. **Elliston, J. E.,** Hypovirulence and chestnut blight research: fighting disease with disease, *J. For.,* 79, 657, 1981.

19. **Elliston, J. E.,** Hypovirulence, in *Advances in Plant Pathology I,* Ingram, D. S. and Williams, P. H., Eds., Academic Press, London, 1982, 1.

20. **Elliston, J. E.,** Characteristics of dsRNA-free and dsRNA-containing strains of *Endothia parasitica* in relation to hypovirulence, *Phytopathology,* 75, 151, 1985.

21. **Elliston, J. E.,** Preliminary evidence for two debilitating cytoplasmic agents in a strain of *Endothia parasitica* from western Michigan, *Phytopathology,* 75, 170, 1985.

22. **Elliston, J. E.,** Further evidence for two cytoplasmic hypovirulence agents in a strain of *Endothia parasitica* from western Michigan, *Phytopathology,* 75, 1405, 1985.

23. **Elliston, J. E., Jaynes, R. A., Day, P. R., and Anagnostakis, S. L.,** A native American hypovirulent strain of *Endothia parasitica, Proc. Am. Phytopathol. Soc.,* 4, 83, 1977.

24. **Fulbright, D. W.,** Effect of eliminating dsRNA in hypovirulent *Endothia parasitica, Phytopathology,* 74, 722, 1984.

25. **Fulbright, D. W.,** A cytoplasmic hypovirulent strain of *Endothia parasitica* without double-stranded RNA (dsRNA), *Phytopathology,* 75, 1328, 1985.

26. **Fulbright, D. W. and Garrod, S. W.,** Double-stranded RNA banding pattern changes in hypovirulent strains of *Endothia parasitica, Phytopathology,* 74, 801, 1984.

27. **Fulbright, D. W. and Weidlich, W. H.,** Interactions of American chestnut and *Endothia parasitica* in Michigan, *Annu. Rep. North. Nut Grow.,* 73, 74, 1982.

28. **Fulbright, D. W., Weidlich, W. H., Haufler, K. Z., Thomas, C. S., and Paul, C. P.,** Chestnut blight and recovering American chestnut trees in Michigan, *Can. J. Bot.,* 61, 3164, 1983.

29. **Garrod, S. W., Fulbright, D. W., and Ravenscroft, A. V.,** Dissemination of virulent and hypovirulent forms of a marked strain of *Endothia parasitica* in Michigan *Phytopathology,* 75, 533, 1985.

30. **Ghabrial, S. A.,** Effects of fungal viruses on their hosts, *Annu. Rev. Phytopathol.,* 18, 441, 1980.

31. **Grente, J.,** Les formes hypovirulentes d'*Endothia parasitica* et les espoirs de lutte contre le chancre du chataignier, *C. R. Acad. Agr. Fr.,* 51, 1033, 1965.

32. **Grente, J. and Sauret, S.,** L' ''hypovirulence exclusive'' phenomene original en pathologie vegetale, *C. R. Acad. Sci., Ser. D.,* 268, 2347, 1969.

33. **Grente, J. and Berthelay-Sauret, S.,** Research carried out in France into diseases of the chestnut tree, in *Proc. Am. Chestnut Symp.,* MacDonald, W. L., Cech, F. C., Luchok, J., and Smith, C., West Virginia University Books, Morgantown, W. Va., 1978, 88.

34. **Griffin, G. J., Hebard, F. V., Wendt, R., and Elkins, J. R.,** Survival of American chestnut trees: evaluation of blight resistance and virulence in *Endothia parasitica, Phytopathology,* 73, 1084, 1983.

35. **Griffin, G. J., Wendt, R. A., and Elkins, J. R.,** Association of hypovirulent *Endothia parasitica* with American chestnut in forest clearcuts and with mites, *Phytopathology,* 74, 804, 1984.

36. **Havir, E. A. and Anagnostakis, S. L.,** Oxalate production by virulent but not by hypovirulent strains of *Endothia parasitica, Physiol. Plant Pathol.,* 23 369, 1983.

37. **Havir, E. A. and Anagnostakis, S. L.,** Oxaloacetate acetylhydrolase activity in virulent but not by hypovirulent strains of *Endothia* (Cryphonectria) *parasitica, Physiol. Plant Pathol.,* 26, 1, 1985.

38. **Heald, F. D., Gardner, M. W., and Studhalter, R. A.,** Air and wind dissemination of ascospores of the chestnut-blight fungus, *J. Agric. Res.,* 2, 405, 1915.

39. **Hepting, G. H.,** Death of the American chestnut, *J. Forest Hist.,* 18, 60, 1974.

40. **Hobbins, D. L. and MacDonald, W. L.,** Interactions between virulent cankers of *Endothia parasitica* and sources of virulent and hypovirulent inoculum on American chestnut, *Phytopathology,* 74, 757, 1984.

41. **Jaynes, R. A. and DePalma, N.,** Effects of blight curing strains on native chestnuts for six years, *Annu. Rep. North. Nut Grow. Assoc.,* 75, 64, 1984.

42. **Jaynes, R. A., and Elliston, J. E.,** Pathogenicity and canker by mixtures of hypovirulent strains of *Endothia parasitica* in American chestnut, *Phytopathology,* 70, 453, 1980.

43. **Jaynes, R. A. and Elliston, J. E.,** Hypovirulent isolates of *Endothia parasitica* associated with large American chestnut trees, *Plant Dis.,* 66, 769, 1982.

44. **Kuhlman, E. G.,** The devastation of American chestnut by blight, in *Proc. Am. Chestnut Symp.,* MacDonald, W. L., Cech, F. C., Luchok, J., and Smith, C., Eds., West Virginia University Books, Morgantown, W. Va., 1978, 1.

45. **Kuhlman, E. G. and Bhattacharyya, H.,** Vegetative compatibility and hypovirulence conversion among naturally occurring isolates of *Cryphonectria parasitica, Phytopathology,* 74, 659, 1984.

46. **Kuhlman, E. G., Bhattacharyya, H., Nash, B. L., Double, M. L., and MacDonald, W. L.,** Identifying hypovirulent isolates of *Cryphonectria parasitica* with broad conversion capacity, *Phytopathology,* 74, 676, 1984.

47. **L'Hostis, B., Hiremath, S. T., Rhoades, R. E., and Ghabrial, S. A.,** Lack of sequence homology between double-stranded RNA from European and American hypovirulent strains of *Endothia parasitica, J. Gen. Virol.,* 66, 351, 1985.

48. **Mittempergher, L.,** The present status of chestnut blight in Italy, in *Proc. Am. Chestnut Symp.,* MacDonald, W. L., Cech, F. C., Luchok, J., and Smith, C., Eds., West Virginia University Books, Morgantown, W. Va., 1978, 34.

49. **MacDonald, W. L. and Double, M. L.,** Frequency of vegetative compatibility types of *Endothia parasitica* in West Virginia, in *Proc. Am. Chestnut Symp.,* MacDonald, W. L., Cech, F. C., Luchok, J., and Smith C., Eds., West Virginia University Books, Morgantown, W. Va., 1978, 103.

50. **Murrill, W. A.,** A serious disease of chestnut, *J. New York Bot. Gard.,* 7, 143, 1906.

51. **Nash, B. L. and Stambaugh, W. J.,** Disease incidence, symptomatology, and vegetative compatibility type distribution of *Endothia parasitica* on oak and chestnut hosts in North Carolina, in *Proc. U.S.D.A.-Forest Service Am. Chestnut Coop. Meet.,* Smith, H. C. and MacDonald, W. L., Eds., West Virginia University Books, Morgantown, W. Va., 1982, 74.

52. **Newhouse, J. R., Hoch, H. C., and MacDonald, W. L.,** The ultrastructure of *Endothia parasitica.* Comparison of a virulent with a hypovirulent isolate, *Can. J. Bot.,* 61, 389, 1983.

53. **Paul, C. P., Estelle, M. A., and Fulbright, D. W.,** Homology among nucleic acids from Michigan hypovirulent strains of *Endothia parasitica, Phytopathology,* 74, 861, 1984.

54. **Paul, C. P. and Fulbright, D. W.,** Homology relationships with Michigan hypovirulent strains of *Endothia parasitica, Phytopathology,* 75, 1325, 1985.

55. Pennsylvania Chestnut Blight Conference, Harrisburg, February 20 and 21, 1912, 252.

56. **Pusey, P. L. and Wilson, C. L.,** Detection of double-stranded RNA in *Ceratocystis ulmi, Phytopathology,* 72, 423, 1982.

57. **Russin, J. S. and Shain, L.,** Initiation and development of cankers caused by virulent and cytoplasmic hypovirulent isolates of the chestnut blight fungus, *Can. J. Bot.,* 62, 2660, 1984.

58. **Russin, J. S., Shain, L., and Nordin, G. L.,** Insects as carriers of virulent and cytoplasmic hypovirulent isolates of the chestnut blight fungus, *J. Econ. Entomol.,* 77, 838, 1984.

59. **Shear, C. L., Stevens, N. E., and Tiller, R. J.,** *Endothia parasitica* and related species, *U.S., Dep. Agric. Bull.,* 380, 82, 1917.

60. **Tipper, D. J. and Bostian, K. A.,** Double-stranded ribonucleic acid killer systems in yeasts, *Microbiol. Rev.,* 48, 125, 1984.

61. **Van Alfen, N. K.,** Biology and potential for disease control of hypovirulence of *Endothia parasitica, Annu. Rev. Phytopathol.,* 20, 349, 1982.

62. **Van Alfen, N. K., Hansen, D. R., Miller, S., and Barley, L.,** Cell-free transmission of hypovirulent phenotype of *Endothia parasitica, Phytopathology,* 74, 833, 1984.

63. **Van Alfen, N. K., Jaynes, R. A., Anagnostakis, S. L., and Day, P. R.,** Chestnut blight: biological control by transmissible hypovirulence in *Endothia parasitica, Science,* 189, 890, 1975.
64. **Van Alfen, N. K., Aloni, H., Hansen, D. R., and Powell, W. A.,** RNA polymerase activity associated with naked dsRNA contained in vesicles of a hypovirulent strain of *Endothia parasitica, Phytopathology,* 75, 1324, 1985.
65. **Willey, R. L.,** Natural dissemination of artificially inoculated strains of *Endothia parasitica,* in *Proc. U.S.D.A.-Forest Service Am. Chestnut Coop. Meet.,* Smith, H. C. and MacDonald, W. L., Eds., West Virginia University Books, Morgantown, W. Va., 1982, 117.
66. **Newhouse, J. R. and MacDonald, W. L.,** unpublished.
67. **Elliston, J. E.,** personal communication.
68. **Endert-Kirkpatrick, E. and Ritchie, D.,** personal communication.
69. **Adams, G., Hammer, S., and Snyder, B.,** personal communication.
70. **Fulbright, D. W.,** unpublished data.
71. **Fulbright, D. W.,** unpublished data.

Chapter 10

BIOLOGICAL CONTROL OF *AGROBACTERIUM* SPECIES

Constance M. E. Garrett

TABLE OF CONTENTS

I. INTRODUCTION

Bacterial diseases of plants are notoriously difficult to control. A wide range of fungicides are available to combat the many, economically important, fungal diseases of plants. The continuing dearth of suitable and effective bactericides for use against the fewer, but sometimes nonetheless economically important, bacterial diseases of crop plants has stimulated a search for alternative methods of control. Although biological control is not a new agricultural technique, it is only in comparatively recent times that serious efforts have been made to exploit the interactions and mutual antagonisms that undoubtedly occur among the complex populations of microorganisms in the relatively stable soil environment. Such interactions have probably been overlooked in the past because there is rarely any notable evidence of their occurrence. However, it is just one such natural interaction that has been successfully exploited in the biological control of crown gall.

A prerequisite for biological control is a knowledge not only of the disease to be controlled but also of the causal organism. Early attempts to understand the biology of crown gall and to effect control of the disease were, to some extent at least, hindered by concentration on the symptoms of the disease rather than on the causal organism. Advances in molecular biology and genetic techniques since 1970 have revolutionized our understanding of crown gall, its causal organism, and consequently our ability to control it.

This chapter looks briefly at the development of biological control for crown gall, discusses some of the successes and failures of its commercial application, and looks to ways in which it might be enhanced and its range extended.

II. CROWN GALL DISEASE

A. Symptoms

Since Smith et al.[66] first reported the bacterial origin of crown gall a vast amount of literature has been accumulated as numerous workers have sought to fathom this fascinating disease. It has been the subject of several excellent reviews,[7,19,33,38,42] so only a brief outline need be given here.

The causal organism, a Gram negative bacterium, *Agrobacterium tumefaciens,* is one of the relatively few soil-inhabiting bacterial plant pathogens. Susceptible plants are invaded only through wounds, usually at the crown or on the root system. These wounds may be natural, originating during seed germination, emergence of lateral roots, from freezing damage, or may be punctures made by feeding of soil fauna. Many wounds are made during normal horticultural practice, e.g., root trimming or breakage preplanting, harvesting of rootstocks from stool or layer beds, or from accidental damage during cultivation. Thus, there are many sites available for ingress of the pathogen. Following infection the plant cells transform into autonomously proliferating tumor cells. The resultant unregulated cell division gives rise to clearly visible and unsightly galls. The young galls are pale, soft, and smooth, but on aging enlarge, darken and, particularly on woody plants, become hard and deeply fissured. Unlike the majority of plant pathogenic bacteria, *A. tumefaciens* has a uniquely vast host range among dicotyledonous plants — over 600 species in 93 families are reported to be susceptible.[15] However, the host range of some strains of the pathogen, e.g., those attacking grapevine, is very limited.[37]

B. The Causal Bacteria

The four species of *Agrobacterium* generally recognized are distinguished from each other on the basis of their pathogenicity. They are *A. tumefaciens, A. rhizogenes,* and *A. rubi* that incite crown gall, hairy root, and cane gall diseases, respectively, and the closely related but nonpathogenic species *A. radiobacter.*[38]

However, since 1970 the unsatisfactory nature of this division of the agrobacteria has become abundantly clear from standard biochemical and physiological tests, studies involving numerical taxonomy, DNA/DNA hybridization studies, and electrophoretograms of soluble proteins. The complexities and current thinking on the relationships of the agrobacteria in the light of these varied studies are admirably reviewed by Kersters and De Ley.[38] The conclusions from all these approaches are that there are three main groups (biovars 1, 2, and 3) within the genus *Agrobacterium* that differ both genetically and phenotypically, but do not correspond to the above species names based on pathogenic behavior. It is the presence, or absence, of certain plasmids which can be lost or acquired with relative ease that determines pathogenicity in these organisms.[75] However, in spite of current awareness of the situation, for the present the old names must remain until a new taxonomy is officially approved and adopted.

C. Mechanism of Action of *A. tumefaciens*

Following the classic experiments of Braun and others in the 1940s, which established that, once initiated, tumor cells could continue to proliferate in the absence of the inciting bacterium, the search was on for the tumor inducing principle (TIP).[7] Since the early 1970s great progress has been made towards elucidating the mechanism of gall production by *A. tumefaciens* as evidenced by the number of papers on the molecular genetics of the pathogen. It is currently believed that virulent cells of the pathogen bind to wounded plant cells and become enmeshed in cellulose fibrils produced by the invading bacteria.[53,67] Virulent *A. tumefaciens* cells contain a large plasmid (mol wt 100 to 150 $\times$ 10^6 Mdaltons) known as pTi which confers on the cell its virulence and tumor-forming ability.[75]

During transformation, part of this pTi (about 5 to 10%) is transferred into the plant cell, by a process as yet undetermined, and becomes integrated with the plant cell DNA causing expression of a new genetic function specified by the T-DNA and not the host cell DNA.[10,72] Thus, the plant cells proliferate to form gall tissue, and synthesize one of a group of unusual amino acids — opines — found only in transformed tumor cells.[5] Particular *A. tumefaciens* strains induce synthesis and catabolism of octopine while other strains have nopaline synthesis and catabolism coded for on pTi; yet others code for agropine. Thus an ecological niche is created for the invading bacterium, since only the inciting strain, or a closely related one, can utilize the particular opine as an energy source and therefore grow around the tumor to the exclusion of other microorganisms.

Loss of pTi is associated with loss of virulence, and acquisition of pTi by an avirulent (*A. radiobacter*) strain from *A. tumefaciens* in the presence of opine confers upon it tumor-inducing ability.[73,74] Other characteristics encoded on pTi include host specificity, binding ability, and sensitivity to the antibiotic agrocin 84. However, sensitivity to the antibiotic is on nopaline type pTi only.

D. Economic Importance

The problems inherent in attempts to estimate losses from plant diseases are well known. With crown gall, however, propagators of planting material have the opportunity to estimate with relative ease the precentage of stock rendered unsalable by the presence of galls; therefore, figures given for crown gall, though not necessarily precise, do give a reasonably clear indication of the level of infection. For example, in Australia up to 80% of peach nursery stock has in the past been rejected because of crown gall.[35,69] The disease has been reported to be prevalent on stone fruits in the U.S., being one of the more serious diseases on crop plants in California and causing overall losses of ca. 10% on the west coast.[62] Even 10 years ago the annual loss from crown gall was estimated to be $2 to 3 million in the U.S.[34] Crown gall is also a problem on stone fruits in Europe.[30,47] Grapevine is also widely attacked[8,39,40] and impaired yield, quality, and growth of up to 20% have been attributed to crown gall infection.[45] Yield and quality of raspberry and of boysenberry have also been badly affected by crown gall infection.[11,54]

The effect of galls on growth and yield of mature trees is difficult to estimate and reports on this aspect are not only sparse but also contradictory.[28] The chief concern over crown gall is infection of nursery or other propagating material.

E. Control Measures

The search for suitable chemicals to control crown gall began soon after nurserymen first became aware of the disease. Numerous substances were tested over many years but little, or only spasmodic, success resulted from these efforts. Soil fumigation was also explored as a means of control, but here again results were inconsistent and probably not economical.[16] A chemotherapeutic control based on a mineral oil emulsion was claimed to be effective but the method of treatment was inappropriate for large scale use.[64] Thus, it was apparent that alternative approaches to control of crown gall were needed if headway was to be made to combat the heavy losses being sustained by nurserymen.

III. BIOLOGICAL CONTROL WITH STRAIN 84 — THEORY

A. History and Development

The much needed breakthrough in the control of the seemingly intractable problem of crown gall came in 1970 when New and Kerr noted that the ratio of pathogenic *A. tumefaciens* to nonpathogenic *A. radiobacter* strains in peach orchards in South Australia was low around healthy trees but high around those with galls.[57] This discovery prompted experiments in which a proportion of nonpathogen to pathogen of $\geq 1:1$ reduced gall infection on greenhouse tomato plants and also on peach seedlings.[57] The strain of *A. radiobacter* used, a particularly effective one and designated strain 84, is now widely distributed and deposited in culture collections. It was fortunate, as Kerr readily acknowledges, that this was among the first of the strains he investigated for antagonistic activity.[35] In further experiments on the incidence of crown gall on peach, inoculation with strain 84 to seeds, roots, or to both seed and roots gave 78, 95, and 99% control, respectively.[32] This was the beginning of the remarkable success story not only of biological control of crown gall but also of the first commercial use of a specific microorganism to control a soil-borne pathogen and the first use of a bacterium to control any plant disease.[56]

Strain 84 is not, however, a universal panacea for all crown gall problems. It is effective against certain strains of the pathogen but not others, and its action is preventive not curative. Nevertheless, in a comparatively short time strain 84 has been widely distributed and used to very good effect.

B. Mechanism of Control by Strain 84

Since the discovery of strain 84 its mechanism of action has been intensively studied and, to a large extent, gradually elucidated. Firstly, only viable cells applied before, together with, or after the pathogen effect control, which precludes the suggestion that the action of strain 84 is that of physical blockage of infection sites.[27,41,52] Kerr and Htay[36] demonstrated that strain 84 produced a bacteriocin to which strains of the pathogen were sensitive, resulting in an inhibition zone around growth of the antagonist on agar plates. Furthermore, strains of the pathogen thus inhibited in vitro could not incite galls on tomato plants when inoculated with strain 84.[36] This antibiotic, agrocin 84, is one of a group of highly specific nucleotide bacteriocins identified as a phosphoramidate of an adenine deoxy-arabinoside.[61] Strain 84 lacks the large Ti plasmid present in virulent *A. tumefaciens* cells, but contains two other plasmids one of which codes for agrocin 84 production and has been designated pAgK84. The other codes for nopaline catabolism and is conjugative.[49] Agrocin 84 inhibits the growth of *A. tumefaciens* strains with a nopaline type of pTi on which agrocin 84 sensitivity is coded[24] by inhibition of nucleic acid and protein synthesis.[14,48] It has been proposed that agrocin 84 is transported into the *A. tumefaciens* cell by the agrocinopine permease system which is particularly efficient in pTi nopaline strains of the pathogen.[71] Evidence has been presented that damage to the outer cell walls of the pathogen, from interference with the synthesis of components, impairs binding to the plant host cells and thereby prevents the essential transfer of bacterial DNA.[47,67]

It should not be assumed that production of agrocin 84 is the sole means whereby control of crown gall is effected. Other strains that produced agrocin 84 had impaired biological control because they failed to grow well at the wound site;[22] while some biological control was obtained by a strain 84 mutant cured of its pAgK84 plasmid, applied to tomato stems 24 hr in advance of the pathogen, apparently by successful competition for attachment sites.[13]

Cognizant that successful biological control of crown gall was dependent on the stability of pathogen sensitivity to agrocin 84, Kerr and Htay[36] were relieved to find that all agrocin 84 resistant mutants arising in the inhibition zones of sensitive strains of the pathogen had simultaneously lost pathogenicity and would not, therefore, pose a threat to breakdown of disease control. This early experience of the Australian workers has not been repeated elsewhere.[70]

IV. BIOLOGICAL CONTROL WITH STRAIN 84 — PRACTICAL APPLICATION

A. Production and Application

The Waite Agricultural Research Institute began supplying growers with agar cultures of strain 84 in 1973.[35] Since then it has been produced in a variety of ways, commercially and experimentally, in many countries.[53] High inoculum concentrations (ca. 10^9 to 10^{10} cfu/mℓ) are usually prepared. Agar cultures, with a shelf life of 6 to 8 weeks without loss of viability, present some distribution problems. If liquid cultures are supplied they need to be separated from the growth medium before use.[53] Cultures impregnated to finely ground peat, based on techniques developed for distribution of *Rhizobium* inocula, have been favored, especially in Australasia, and the 6-month shelf life claimed for such preparations is confirmed.[43,70] Strain 84 has also been maintained in 0.25% carboxymethyl cellulose for 10 months, but this leads to physiologically retarded cells that need to be used at higher than routine concentrations for effective biological control.[53] The practicality of using freeze-dried cultures, having a long period of viability, has been under investigation.[3]

Strain 84 can be a cheap and effective control of crown gall, but it is still not yet available to growers in all countries because legal requirements for registration and use of a living bacterium have either not been explored or not been granted. There was no problem in the registration of peat-based cultures in New Zealand because of parallels with *Rhizobium* inoculants.[76] Initial opposition by the U.S. Environmental Protection Agency was overcome in 1979 when toxicological data met safety requirements, and permission was granted for the use of strain 84 on immature fruit trees.[77] The generally increased interest in nonpolluting and environmentally acceptable biological controls should facilitate registration of antagonistic organisms in the future, and previously fought battles for approval of strain 84 must surely have paved the way for its even more widespread registration. However, in countries where the disease is not a widespread and recurrent problem, commercial firms are unconvinced that production of strain 84 would be economic.

In whatever formulation or preparation growers receive strain 84 they need to be cognizant of the strictures imposed by the use of a living organism, if they are to achieve the desired effect. The use of equipment or water contaminated with chemicals likely to be harmful to the bacteria, or exposure of the inoculum, or plants immediately after treatment, to bright sunlight or extreme temperatures, may not only reduce the viability of the antagonist to ineffective levels but could also jeopardize future control (see below).

Even where pathogenic strains of *A. tumefaciens* are known to be sensitive to agrocin 84 it is essential to apply the antagonist at the right place and right time to effect control. The opportunities for establishing strain 84 on the vulnerable areas of the host are few, i.e., when sowing seeds, or when transplanting seedlings, cuttings, or young rootstocks. Dipping seeds or plant material in a suspension of the antagonist is preferable. A decrease in control is reported if strain 84 is applied to soil into which planting is to take place, or if it is applied to seeds on the open ground or poured on to emerging seedlings and the surrounding soil.[43,46]

In attempting to assess the value of strain 84 as a biological control agent it is necessary to distinguish between control of natural field infestation, which is usually very good, and those experiments in which the pathogen is applied to all plants and strain 84 is applied before, with, or after the pathogen or not at all.

B. Successful Controls

Under natural field conditions the population of pathogenic agrobacteria will be relatively low. Where such populations are composed of agrocin 84–sensitive strains the application of a high (10^8 cfu/mℓ) dose of strain 84 should readily eliminate crown gall infection. The already impressive list of successes achieved with strain 84 is a testimony to the efficacy of the treatment, especially on stone fruits and roses.

Probably the most widespread use of strain 84 has been in Australia where, since 1973, stone fruit and rose growers have given their nursery stock preplanting dips of strain 84, and thereby achieved nearly complete control of a disease that was formerly a serious problem.[35] Crown gall on peaches, nectarines, and roses is now rarely seen in New Zealand thanks to the widespread use of this method of biological control.[21] In Canada and South Africa, strain 84, as a presowing dip for peach seed, has virtually eliminated the incidence of crown gall on this host.[17,46] Levels of crown gall, commonly as high as 30% on peach and almond seedlings planted into heavily infested land in Greece, have been reduced to almost nil.[60] In Hungary and Switzerland crown gall on peach, apricot, and cherry has been controlled by applications of strain 84,[30] while excellent control of the disease on rose cuttings has been reported from Italy.[26]

The variable levels of infestation with *A. tumefaciens* and its uneven distribution in a field soil make experimentation difficult, and necessitate a high degree of replication. Many researchers have, therefore, assessed the efficacy of biological control by strain 84 by applying it before, with, or after treatment with the pathogen at a ratio of $\geq 1:1$. Under these conditions, where the antagonist is challenged with much higher populations of the pathogen than it would encounter under natural conditions, control can still be very effective and reach levels of 95% or more.

Good control of crown gall under such experimental conditions has been reported with strain 84 on a variety of stone fruit hosts.[4,20,25,29,52,65] Moore[52] found that application of strain 84 to wounds several hours in advance of inoculation with the pathogen gave enhanced control. In practice this will often be done when root-pruned cuttings are treated, but their planting is delayed due to inclement conditions. Crown gall, commonly a problem on rose rootstocks in Italy, could be substantially reduced (70 to 100 %) by treatment with strain 84 before dipping in the pathogen.[63] Virtual freedom from crown gall was obtained when cuttings of willow (*Salix purpurea*) were dipped in strain 84 prior to planting out.[68] Results have been generally less good when attempts have been made to control crown gall on herbaceous hosts or those subject to production of aerial galls, e.g., grapevine and cane fruits. On grapevine the pathogen is known to be systemic.[8] Variable results for biological control of crown gall on chrysanthemum have been obtained. Miller and Miller[50] claimed to have good control when plants were injected with virulent *A. tumefaciens* 24 hr after inoculation with strain 84. In Europe encouraging results have been reported using strain

84 on chrysanthemum and pelargonium,[9,25] but the experience of other workers, notably in Australasia and Italy, suggests that strain 84 does not give a satisfactory level of control of crown gall on many herbaceous hosts, including chrysanthemum[3,35,58]

Another host on which strain 84 has failed to control crown gall is grapevine. This is attributed to the fact that the biovar 3 strains of the pathogen that cause the disease on grapevines are insensitive to agrocin 84.[35,37] However, sensitivity to the antagonist was recently reported when a modified Stonier's medium was used. Furthermore, gall production by the biovar 3 strains examined, pathogenic to datura, tomato, and some other hosts too, was suppressed if strain 84 was applied to stems of the former hosts several hours in advance of the pathogen.[18]

Similarly, application of strain 84 to cherry and tomato plants 24 hr in advance of apparently agrocin 84-resistant strains of the pathogen was reported to have controlled gall formation. Subsequently these strains were found to be sensitive to agrocin 84 in vitro when a modified test medium was used.[12]

C. Ineffective Controls

Even where agrocin 84-sensitive strains are known to be present, biological control of crown gall with strain 84 is not always successful. Field control will be reduced when a significant proportion of the population of the pathogen is composed of strains resistant to the action of the antagonist, whether naturally occurring or developed by mutations occurring after application of strain 84. It has been suggested that the almost complete success of the biological control in Australasian stone-fruit and rose nurseries is due to the uniformity of the pathogen in these soils, whereas in other places the populations of *A. tumefaciens* are more variable.[53] A survey of Spanish crown gall hosts and localities, for example, revealed that about 20% of isolates of *A. tumefaciens* sampled were resistant to strain 84.[44]

Several strains of the pathogen from peaches in Greece were resistant to strain 84 and thus not controlled by it. The poor control of crown gall from the use of strain 84 at low concentrations (10^5 or 10^6 cfu/mℓ) on willow cuttings was attributed to the 2% of field strains of the pathogen that in vitro tests had shown were insensitive to strain 84.[68] Alternatively, the low concentrations of strain 84 used could have permitted development of transconjugants having agrocin resistance and pathogenicity.

Alconero[1] claimed that a correlation existed between the degree of sensitivity to agrocin 84 and the degree of biological control that could be achieved. He suggested that the limited success of biological control reported by Miller et al.[51] was in part due to the fact that 10% of the strains were only weakly sensitive to the antagonist and thus higher ratios of antagonist to pathogen would be needed. Similarly, the 15% of the population that were only moderately sensitive to agrocin 84 were implicated in the rather poor control of crown gall of peaches.[20]

Although strain 84 has given good control of crown gall on stone fruits it almost universally fails to control the disease on pome fruits.[3,30] On apple, as on grapevine, the particular form of the pathogen inciting disease appears to be rather host specific.[2] Not only is strain 84 ineffective in controlling crown gall in apple but it can actually enhance the incidence of disease. Thus, on M4 rootstocks the incidence of crown gall on strain 84-treated stocks compared to controls was increased from ca. 6% to ca. 14% when worked with Idared, and from ca. 6.0% to 20.5% when worked with Starking, but remained at ca. 2.5% when worked with Jonathan.[30] This result emphasizes that it is important to test the efficacy of strain 84 on a particular crop before attempting widescale use, and also draws attention to the influence of scion variety on the susceptibility of the rootstock to crown gall, as has also been observed in cherry.[28] The mechanism of such an interaction is unknown. A report that M9 apple rootstocks treated with strain 84 had double the amount of galls on controls must be discounted because there was a strong possibility of latent crown gall before treatment. In all experiments it is essential to be certain that plants are totally free from infection initially. Moore[53] reported a site effect on the incidence of crown gall on apple seedlings at three nurseries in central Washington. He demonstrated that at the two sites most conducive to the disease, approximately a three-fold increase in the percentage of galled seedlings followed treatment with a mixture of strains of the pathogen together with strain 84, compared with those treated with the pathogens alone.

Why strain 84 fails to control crown gall on pome fruits, but is effective on stone fruits, has been attributed by Cooksey and Moore[12] to root exudates from pome fruits affecting the sensitivity of strains of the pathogen to agrocin 84 in a manner similar to that which allows demonstration of sensitivity in vitro on some media and not on others. It is possible that a similar explanation could be advanced for failure on some other hosts, but since apple and grapevine strains of the pathogen appear to be host adapted forms rather than of wide host range such as the stone-fruit pathogens, some other mechanism may be involved.

D. Breakdown in Control

In Australia where strain 84 has been used to control crown gall infection since 1972 there have not yet been any reports that it is losing its effectiveness.[35] Although the in vitro mutants that Htay and Kerr tested were both agrocin 84 resistant and avirulent,[32] Süle found a high proportion of mutants derived in vitro from agrocin 84–sensitive strains were resistant to the antagonist and also pathogenic to sunflower. Thus they were not controlled by strain 84 on inoculation to tomato stems.[70] The first report of reduced efficacy in the field came in 1979 when Panagopoulos et al.[59] noted a rapid breakdown in control of crown gall on peach seedlings inoculated with a 1:1 ratio of pathogen and antagonist. From the resultant galls, 16.5% of the agrobacteria isolated were pathogenic, agrocin 84 producer strains and thus were insensitive to agrocin 84. They were considered to pose a considerable threat to biological control.

In subsequent laboratory experiments strain 84 crossed with virulent strains of the pathogen yielded transconjugants resulting from transfer of the agrocin producer plasmid to the pathogen or acquisition of Ti plasmid by strain 84 cells.[23] In other experiments more than 100 transconjugants were obtained both in vitro and in planta after exposure of two strains of *A. tumefaciens* to strain 84. About 50% from one parent and 5% from the other were virulent (nopaline pTi) and either resistant or semiresistant to strain 84.[71] On the other hand the stability of pTi was apparently low when a mutation rate of 10^{-4} of three *A. tumefaciens* strains yielded, on exposure to agrocin 84, a majority of agrocin 84 resistant mutants (68, 28, and 80% of mutants from the three strains) that were nonpathogenic, while few, but nevertheless a significant number, had retained pTi and the pathogenicity of the parent isolates. One *A. rhizogenes* strain had a mutation rate of 10^{-3}, all mutants being agrocin 84-resistant and nonpathogenic.[13] Such plasmid transfer occurs only in the presence of nopaline. Therefore, where a sufficiently high concentration of strain 84 is applied in the field to completely inhibit agrocin 84-sensitive (pTi nopaline) strains, gall tissue and its associated opine is not produced. Therefore undesirable mutants cannot arise to reduce the efficacy of biological control. It has already been suggested that the failure of the lower concentrations of strain 84 used against willow crown gall could have been due to the formation of undesirable agrocin-resistant conjugants in this way.

V. ALTERNATIVES TO STRAIN 84

A. Natural

Ever since it was realized that strain 84 was effective against pTi nopaline strains of *A. tumefaciens* only, a search has been conducted to find equally effective antagonists of pTi octopine and biovar 3 strains of the pathogen. Several workers have tested the biological potential of hundreds of other agrocin-producing strains of agrobacteria but none has proved sufficiently effective in in planta tests to warrant their further investigation as an alternative to strain 84.[27,37,56]

One exception is *A. tumefaciens* strain D286, a biovar 1 strain from a eucalyptus gall, that produces agrocin D286 having a broader host range than strain 84. It inhibits growth of nopaline, octopine, and agropine metabolizing strains of the pathogen, and when reinoculated into plants was found to have lost its pathogenicity.[31] Its broader host range, and faster growth rate than strain 84 indicate that it is potentially a valuable additional biological control agent for crown gall. It does not seem to have been exploited.

The potential of soil fungi as antagonists has been investigated and several were antagonistic to agrobacteria in vitro, but this was not matched in in vivo tests. Some *Penicillium* and *Aspergillus* species and also a *Bacillus* were able to reduce the level of crown gall when inoculated to test plants in advance of the pathogen, but nonetheless their activity was inferior to that of strain 84.[12] However, in field tests against low natural populations of the pathogen, a *Penicillium* and a *Bacillus* effectively controlled crown gall indicating that competitors may play an important role in the natural suppression of crown gall, as well as fluorescent pseudomonads and actinomycetes. The use of some of these antagonists in pecan orchards, where only a third of the biovar 2 population of *A. tumefaciens* was sensitive to strain 84, has been considered.[6] Following an observation that a sludge-treated soil in which raspberries were grown led to less crown gall infection than on comparable nontreated soils Moore et al.[55] demonstrated the effect was due to antagonistic organisms and not to toxic trace elements in the sludge inactivating the pathogen. It was not clear whether sludge components affected plant growth and/or sensitivity to crown gall. Although greenhouse experiments have given disappointing results the phenomenon is regarded as worthy of exploitation.[78]

The hypothesis that increase in crown gall following soil fumigation may be due to elimination of competitive, antagonistic organisms[12] would seem to be a valid one, and a beneficial effect should result from early recolonization of fumigated soils with selected bacterial or fungal antagonists of *Agrobacterium*.

B. Engineered

The unique role of strain 84 in biological control of crown gall has prompted investigations to try to manipulate the strain to produce one of increased range or activity. Transfer of the agrocin producer plasmid (pAg84) to avirulent recipients resulted in about 10% of them becoming agrocin 84 producers but they were ineffective in crown gall control on tomato stems, probably due at least in part to poor growth at the site of action.[22]

It has been argued that to select from strain 84 a mutant with a deficient plasmid transfer system might significantly reduce the risk of breakdown in control that could conceivably result from development of undesirable transconjugants in the field.[59,60]

A recombinant strain, designated 0341, more antagonistic than strain 84, lacks conjugative ability and is thus less liable to transfer the pAg84 plasmid.[60] This potentially useful antagonist does not however appear to have been exploited yet.

VI. PROSPECTS FOR THE FUTURE

The success of strain 84 as an antagonist has stimulated research for other organisms to extend the range of agrobacteria that can be controlled biologically and to reduce risks of breakdown in such control. One of the major difficulties is to obtain organisms that not only meet the above requirements but also are as effective in vivo as in vitro tests indicate that they should be. Reasons for past failures in this respect may be due to the inability to establish or grow well in the ecological niche to which they must be introduced, poor productivity of the effective antibiotic in this environment, or suppression by other plant or soil organisms. Careful selection of organisms, bearing in mind these criteria, should result in a wider range of successful antagonists. Great attention should be paid to those already well adapted to the ecological niche in which they must operate. A "cocktail" of several organisms might be more effective than reliance upon a single strain.

Although the emphasis so far has been on agrocin producing strains this is not the only attribute of a successful antagonist. At least one mutant of strain 84 cured of its pAg84 was found to have some effect on crown gall, probably because of successful competition for attachment sites.[12]

Thus, the search must continue for organisms which can produce an antibiotic, successfully compete in the site of action, and have a minimal risk of inducing a breakdown in the biological control for which they have been selected. The prospects for extending biological control of crown gall on hosts or biovars of the pathogen not yet subject to it are encouraging and merit further research.

REFERENCES

1. **Alconero, R.,** Crown gall of peaches from Maryland, South Carolina and Tennessee and problems with biological control, *Plant Dis.,* 64, 835, 1980
2. **Anderson, A. R. and Moore. L. W.,** Host specificity in the genus *Agrobacterium, Phytopathology,* 69, 320, 1979.
3. **Bazzi, C.,** Biological control of crown gall in Italy, International Workshop on Crown Gall, Swiss Federal Research Station, Wadenswil, Switzerland, 1983,1.
4. **Bazzi, C. and Mazzucchi, U.,** Biological control of crown gall on sweet cherry, myrobalan and peach seedlings in Italy, in *Proc. 4th Int. Conf. Plant Pathogenic Bacteria, Angers, 1978,* Station de Pathologie Vegetale et Phytobacteriologie, Ed., Institut National de la Recherche Agronomique, 1979, 251.
5. **Bomhoff, G., Klapwijk, P. M., Kester, C. M., and Schilperoort, R. A.,** Octopine and nopaline synthesis and breakdown genetically controlled by a plasmid of *Agrobacterium tumefaciens, Mol. Gen. Genet.,* 145, 177, 1976.
6. **Bouzar, H., Moore, L. W., and Schaad, N. W.,** Crown gall of pecan: a survey of *Agrobacterium* strains and potential for biological control in Georgia, *Plant Dis.,* 67, 310, 1983.
7. **Braun, A. C.,** Tumor inception and development in the crown gall disease, *Annu. Rev. Plant Physiol.,* 13, 533, 1962.
8. **Burr, T. J. and Katz, B. H.,** Isolation of *Agrobacterium tumefaciens* biovar 3 from grapevine galls and sap and from vineyard soil, *Phytopathology,* 73, 163, 1983.
9. **Carta, C., Fiori, M., and Franceschini, A.,** Prova di lotta biologica al tumore batterico del crisantemo (*Chrysanthemum morifolium* Ramat) in Sardegna, *Phytopathol. Mediterr.,* 22, 217, 1983.
10. **Chilton, M-D., Drummond, M. H., Merlo, D. J., Sciaky, D., Montoya, A. L., Gordon, M. P., and Nester, E. W.,** Stable incorporation of plasmid DNA into higher plant cells: the molecular basis of crown gall tumorigenesis, *Cell,* 11, 263, 1977.
11. **Converse, R. H.,** Diseases of raspberries and erect and trailing blackberries, *U.S. Dept. Agric. Handbo.,* No. 310, 1, 1966.
12. **Cooksey, D. A. and Moore, L. W.,** Biological control of crown gall with fungal and bacterial antagonists, *Phytopathology,* 70, 506, 1980.
13. **Cooksey, D. A. and Moore, L. W.,** High frequency spontaneous mutation to agrocin 84 resistance in *Agrobacterium tumefaciens* and *A. rhizogenes, Physiol. Plant Pathol.,* 20, 129, 1982.
14. **Das, P. K., Basu, M., and Chatterjee, G. C.,** Studies on the mode of action of agrocin 84, *J. Antibiot.,* 31, 490, 1978.
15. **De Cleene, M. and De Ley, J.,** The host range of crown gall, *Bot. Rev.,* 42, 389, 1976.
16. **Deep, I. W., McNeilan, R. A., and MacSwan, I. C.,** Soil fumigants tested for control of crown gall, *Plant Dis. Rep.,* 52, 102, 1968.
17. **Dhanvantari, B. N.,** Biological control of crown gall of peach in Southwestern Ontario, *Plant Dis. Rep.,* 60, 549, 1976.
18. **Dhanvantari, B. N.,** Etiology of grape crown gall in Ontario, *Can. J. Bot.,* 61, 2641, 1983.
19. **Drummond, M.,** Crown gall disease, *Nature (London),* 281, 343, 1979.
20. **Du Plessis, H. J., Hattingh, M. J., and Vanvuuren, H. J. J.,** Biological control of crown gall in South Africa by *Agrobacterium radiobacter* strain 84, *Plant Dis.,* 69, 302, 1985.

21. **Dye, D. W., Kemp, W. J., Amos, M. J., and Parker, W. C.,** Crown gall in roses can be controlled, *Commercial Horticulture,* 7, 5, 1975.
22. **Ellis, J. G. and Kerr, A.,** Developing biological control agents for soil borne pathogens, *Proc. 4th Int. Conf. Plant Pathogenic Bacteria, Angers, 1978,* Station de Pathologie Vegetale et Phytobacteriologie, Ed., Institut National de la Recherche Agronomique, 1979, 245.
23. **Ellis, J. G. and Kerr, A.,** Transfer of agrocin 84 production from strain 84 to pathogenic recipients: a comment on the previous paper, in *Soil-borne Plant Pathogens,* Schippers, B. and Gams, W., Eds., Academic Press, London, 1979, 579.
24. **Engler, G., Hosters, M., Van Montagu, M., Schell, J., Hernalsteens, J. P., and Schilperoort, R. A.,** Agrocin 84 sensitivity: a plasmid determined property in *Agrobacterium tumefaciens, Mol. Gen. Genet.,* 138, 145, 1975.
25. **Faivre-Amiot, A., Roux, J., and Faivre, M. L.,** Essai de lutte biologique contra la galle du collet (crown gall) a l'aide d'une souche d'*Agrobacterium* non pathogene: *Agrobacterium radiobacter* var. *radiobacter* (souche 84 de Kerr), *Phytiatr. Phytopharm.,* 28, 203, 1979.
26. **Garibaldi, A. and Raimondi, M.,** (Biological control of bacterial canker of rose.) L'impiego della lotta biologica contro il tumore batterico della rosa, *Atti Manifestazioni in versilia, vioreggio,* p. 184, 1977.
27. **Garrett, C. M. E.,** Biological control of crown gall in cherry rootstock propagation, *Ann. Appl. Biol.,* 91, 221, 1979.
28. **Garrett, C. M. E.,** The effect of crown gall on growth of cherry trees, *Plant Pathol.,* 36, 339, 1987.
29. **Garrett, C. M. E. and Lemmon, V. M.,** Crown gall (*Agrobacterium tumefaciens*) Biological control, *East Malling Res. Stn. Rep. 1978.* Maidstone, England, 1979, 84.
30. **Grimm, R. and Süle, S.,** Control of crown gall (*Agrobacterium tumefaciens* Smith and Townsend) in nurseries, in *Proc. 5th Int. Conf. Plant Pathogenic Bacteria, Cali, Columbia, 1981.* Lozano, J. C., Ed., Centro International Agricultura Tropical, Cali, Columbia, 1982, 531.
31. **Hendson, M., Askjaer, L., Thomson, J. A., and Van Montagu, M.,** Broad-host-range agrocin of *Agrobacterium tumefaciens, Appl. Environ. Microbiol.,* 45, 1526, 1983.
32. **Htay, K. and Kerr, A.,** Biological control of crown gall: seed and root inoculation, *J. Appl. Bacteriol.,* 37, 525, 1974.
33. **Kado, C. I.,** The tumor-inducing substance of *Agrobacterium tumefaciens, Annu. Rev. Phytopathol.* 14, 265, 1976.
34. **Kennedy, B. W. and Alcorn, S. M.,** Estimates of U.S. crop losses due to procaryote plant pathogens, *Plant Dis.,* 64, 674, 1980.
35. **Kerr, A.,** Biological control of crown gall through production of agrocin 84, *Plant Dis.,* 64, 25, 1980.
36. **Kerr, A. and Htay, K.,** Biological control of crown gall through bacteriocin production, *Physiol. Plant Pathol.,* 4, 37, 1974.
37. **Kerr, A. and Panagopoulos, C. G.,** Biotypes of *Agrobacterium radiobacter* var. *tumefaciens* and their biological control, *Phytopathol. Z.,* 90, 172, 1977.
38. **Kersters, K. and De Ley, J.,** Genus III. *Agrobacterium* Conn 1942, in *Bergey's Manual of Systematic Bacteriology,* Vol. 1, Krieg, N. R. and Holt, J. G., Eds., Williams & Wilkins, Baltimore, 1984, 244.
39. **Lehoczky, J.,** Spread of *Agrobacterium tumefaciens* in the vessels of the grapevine, after natural infection, *Phytopathol. Z.,* 63, 239, 1968.
40. **Lehoczky, J.,** Further evidences concerning the systemic spreading of *Agrobacterium tumefaciens* in the vascular system of grapevine, *Vitis,* 10, 215, 1971.
41. **Lippincott, B. B. and Lippincott, J. A.,** Bacterial attachment to a specific wound site as an essential stage in tumor initiation by *Agrobacterium tumefaciens, J. Bacteriol.,* 97, 620, 1969.
42. **Lippincott, J. A. and Lippincott, B. B.,** The genus *Agrobacterium* and plant tumorigenesis, *Annu. Rev. Microbiol.,* 29, 377, 1975.
43. **Lopez, M. M., Miro, M., Orive, R., Temprano, F., and Poli, M.,** Biological control of crown gall of rose in Spain, *Proc. 5th Int. Conf. Plant Pathogenic Bacteria, Cali, Columbia, 1981,* Lozano, J. C., Ed., Centro International de Agricultura Tropical, Cali, Columbia, 1982, 538.
44. **Lopez, M. M., Miro, M., Gorris, M. T., Salcedo, C. I., Temprano, F., and Orive, R. J.,** Comparative efficiency of inoculation treatments with *Agrobacterium radiobacter* pv. *radiobacter* K84 against sensitive and resistant agrocin 84 strains of *Agrobacterium radiobacter* pv. *tumefaciens,* International Workshop on Crown Gall, Swiss Federal Research Station, Wadenswil, Switzerland, 1983, 43.
45. **Malenin, I.,** (The effect of bacterial canker of the vine variety Bolgar.) Vliyanie na bakterialniya vak vurkhu sort Bolgar, *Gradinar. Lozar. Nauka,* 8, 101, 1971.
46. **Matthee, F. N., Thomas, A. C., and Du Plessis, H. J.,** Biological control of crown gall, *Deciduous Fruit Grower,* 27, 303, 1977.
47. **Matthysse, A. G., Holmes, K. V., and Gurlitz, R. H. G.,** Elaboration of cellulose fibrils by *Agrobacterium tumefaciens* during attachment to carrot cells, *J. Bacteriol.,* 145, 583, 1981.
48. **McCardell, B. A. and Pootjes, C. F.,** Chemical nature of agrocin 84 and its effect on a virulent strain of *Agrobacterium tumefaciens, Antimicrob. Agents Chemother.,* 10, 498, 1976.

49. **Merlo, D. J. and Nester, E. W.**, Plasmids in avirulent strains of *Agrobacterium, J. Bacteriol.,* 129, 76, 1977.

50. **Miller, H. N. and Miller, J. W.**, Control of crown gall on chrysanthemum with a non-pathogenic bacterium and with selected antibiotics, *Proc. Fl. State Hortic. Soc.,* 89, 301, 1977.

51. **Miller, R. W., Brittain, J. A., and Watson, T.**, Biological control of crown gall on peach, *Fruit South,* 3, 30, 1979.

52. **Moore, L. W.**, Prevention of crown gall on *Prunus* roots by bacterial antagonists, *Phytopathology,* 67, 139, 1977.

53. **Moore, L. W.**, Practical use and success of *Agrobacterium radiobacter* strain 84 for crown gall control, in *Soil-borne Plant Pathogens,* Schippers, B. and Gams, W. Eds., Academic Press, London, 1979, 553.

54. **Moore, L. W.**, Controlling crown gall with biological antagonists, *Am. Nurseryman,* 151, 40, 1980.

55. **Moore, L. W., Volk, V. V., and Morris, L.**, Control of crown gall disease with sewage sludge, International Workshop on Crown Gall, Swiss Federal Research Station, Wadenswil, Switzerland, 1983, 59.

56. **Moore, L. W. and Warren, G.**, *Agrobacterium radiobacter* strain 84 and biological control of crown gall, *Annu. Rev. Phytopathol.,* 17, 163, 1979.

57. **New, P. B. and Kerr, A.**, Biological control of crown gall: field measurements and glasshouse experiments, *J. Appl. Bacteriol.,* 35, 279, 1972.

58. **New, P. B. and Milne, K. S.**, Etiology and control of crown gall on potted chrysanthemums, *N. Z. J. Exp. Agric.,* 4, 109, 1975.

59. **Panagopoulos, C. G., Psallidas, P. G., and Alivizatos, A. S.**, Evidence of a breakdown in the effectiveness of biological control of crown gall, in *Soil-borne Plant Pathogens,* Schippers, B. and Gams, W., Eds., Academic Press, London, 1979, 569.

60. **Panagopoulos, C. G., Psallidas, P. G., and Stylianidis, D. C.**, Current work on the effectiveness of biocontrol of crown gall, International Workshop on Crown Gall, Swiss Federal Research Station, Wadenswil, Switzerland, 1983, 67.

61. **Roberts, W. P., Tate, M. E., and Kerr, A.**, Agrocin 84 is a 6-N-phosphoramidate of an adenine nucleotide analogue, *Nature (London),* 265, 379, 1977.

62. **Ross, N., Schroth, M. N., Sanborn, R., O'Reilly, H. J., and Thompson, J. P.**, Reducing loss from crown gall disease, *Calif. Agric. Exp. Stn. Bull.,* p. 845, 1970.

63. **Rumine, P. and Comucci, A.**, Applications of biological control of bacterial canker on rose rootstocks, *Inf. Fitopatol.,* 34, 39, 1984.

64. **Schroth, M. N. and Hildebrand, D. C.**, A chemotherapeutic treatment for selectively eradicating crown gall and olive knot neoplasms, *Phytopathology,* 58, 848, 1968.

65. **Schroth, M. N. and Moller, W. J.**, Crown gall controlled in the field with a non-pathogenic bacterium, *Plant Dis. Rep.,* 60, 275, 1976.

66. **Smith, E. F., Brown, N. A., and Townsend, C. O.**, Crown gall of plants: its cause and remedy, *U.S. Dept. Agric. Bur. Plant Ind. Bull.,* 213, 1, 1911.

67. **Smith, V. A. and Hindley, J.**, Effects of agrocin 84 on attachment of *Agrobacterium tumefaciens* to cultured tobacco cells, *Nature (London),* 276, 498, 1978.

68. **Spiers, A. G.**, Biological control of *Agrobacterium* species on Salix, *N. Z. J. Agric. Res.,* 23, 139, 1980.

69. **Stubbs, L. L.**, Plant pathology in Australia, *Rev. Plant Pathol.,* 50, 461, 1971.

70. **Süle, S.**, Pathogenicity and agrocin 84 resistance of *Agrobacterium tumefaciens, Proc. 4th Int. Conf. Plant Pathogenic Bacteria, Angers, 1978,* Station de Pathologie Vegetale et Phytobacteriologie, Ed., Institut National de la Recherche Agronomique, 1979, 255.

71. **Süle, S. and Kado, C. I.**, Agrocin resistance in virulent derivatives of *Agrobacterium tumefaciens* harbouring the pTi plasmid, *Physiol. Plant Pathol.,* 17, 347, 1980.

72. **Thomashow, M. F., Nutter, R., Montoya, A. L., Gordon, M. P., and Nester, E. W.**, Integration and organization of Ti-plasmid sequences in crown gall tumors, *Cell,* 19, 729, 1980.

73. **Van Larebeke, N., Genetello, C. H., Schell, J., Schilperoort, R. A., Hermans, A. K., Hernalsteens, J. A., and Van Montagu, M.**, Acquisition of tumor-inducing ability by non-pathogenic agrobacteria as a result of plasmid transfer, *Nature (London),* 255, 742, 1975.

74. **Watson, B., Currier, T., Gordon, M. P., Chilton, M-D., and Nester, E. W.**, Plasmid required for virulence of *Agrobacterium tumefaciens. J. Bacteriol.,* 123, 255, 1975.

75. **Zaenen, I., Van Larebeke, N., Teuchy, H., Van Montagu, M. and Schell, J.**, Supercoiled circular DNA in crown gall inducing *Agrobacterium* strains, *J. Mol. Biol.,* 86, 109, 1974.

76. **Dye, D. W.**, D.S.I.R., Aukland, New Zealand, personal communication.

77. **Moore, L. W.**, Oregon State University, Corvallis, Ore., personal communication.

78. **Moore, L. W.**, Oregon State University, Corvallis, Ore., personal communication.

Chapter 11

BIOLOGICAL CONTROL OF DISEASES OF FRUITS

Wojciech Janisiewicz

TABLE OF CONTENTS

I. INTRODUCTION

Increased interest in biological control in the last decade has been accompanied by many unsuccessful attempts at transferring potentially effective biocontrol systems from the laboratory to the field. However, this research assumes greater importance every year. The primary reasons for continuing studies in this area have been a few examples of successful biocontrol, increased failure of current pesticides, higher cost of development and greater difficulties associated with finding new pesticides, many new "iatrogenic" diseases (" . . . diseases which result from or are increased in severity by the use of a specific crop protection chemical . . . "),[34,39,43,62] and finally, new opportunities opened up by genetic engineering.[7,50,53] Multiple findings of pesticide residues in different commodities, as well as toxins produced by fruit pathogens, have stimulated public demand for fruits and vegetables free of disease and hazardous chemical residues.

All of the foregoing creates a positive climate for the development of alternative methods for control of fruit diseases. Biological control is one such alternative.

Biocontrol of diseases on above ground parts presents a major challenge, and has been successful only in a few cases.[8,22,66,51] Environmental factors play a major role in determining its success. Control of these factors is much simpler in naturally buffered soil environments than on plant parts above ground, unless the latter environment can be controlled. Such controlled conditions can exist in greenhouses and storage rooms for fruit and other commodities. Therefore, a controlled environment approach seems to be a sensible step toward shifting biocontrol from the laboratory to the "outside environment" where it can be implemented, tested for its feasibility, and finally commercialized. Disease control on stored fruit presents a special opportunity for testing the effectiveness of biocontrol agents in the "outside environment" where critical factors can be controlled.

In this chapter, I will try to describe pioneering work on biocontrol of diseases of harvested fruit, and some attempts to control fruit diseases in the field with the intent of extending its effect after harvest.

II. FRUIT AS AN ENVIRONMENT FOR MICROORGANISMS

To understand how to use biological agents against fruit pathogens, one needs to know how fruit functions as a site for disease development. The presence of a pathogen on a host alone is not sufficient to cause disease; biotic and abiotic factors must be favorable. Stresses to which biocontrol organisms are exposed during their lives in association with plants have a detrimental effect on their survival and the outcome of biocontrol by resident or introduced antagonists. Here, I will give only a general overview of the most critical points in disease development. For a more detailed description of different types of infection and fruit environments, the reader is referred to other reviews.[25,30]

A. Fruit as a Source of Nutrients for Microorganisms

Fruits have a moist, nutrient-rich environment in which resistance to disease decreases as maturation progresses. Fruit surfaces also provide a very good food base for epiphytic microorganisms, both saprophytic and parasitic. In addition to nutrients coming from regular plant leakage, fruit surfaces may contain outside deposits of pollen, organic debris, or honeydew, which supply significant amounts of nutrients for epiphytic microorganisms.[12] Changes in the nutritional status of fruit as the season progresses cause changes in microbial composition on the fruit surface.[23,24,64,75,76] Fruit pH could also determine which group of organisms (bacteria or fungi) would predominate. Therefore, it is not surprising that when other environmental factors such as temperature and moisture are favorable, microorganisms may multiply rapidly on the fruit surface.

B. Fruit as an Infection Site

Infection by fruit pathogens may occur as early as the blossom stage, e.g., latent infection of strawberry[55] and apple[71,72] by *Botrytis cinerea*, or stone fruits by *Monilinia fructicola*.[67] These pathogens can infect through flower petals, mainly in the senescent stage. After germination and formation of infection structures, the further development of infection is delayed until the fruit ripens, at which time the disease progresses rapidly. Colletotrichum diseases on many tropical fruits (e.g., citrus, banana, papaya) also arise from latent infections. Following spore germination and piercing of the cuticle, a tough appressorium is formed which strongly adheres to the surface. The appressorium may remain dormant for a few months until a more favorable environment (namely fruit maturation) for further infection occurs.

Lenticel infection with *Gleosporium album* on apples is another example of infection which, after initial development, ceases until the fruit ripens in storage. Lenticel and stomata infections have increased in importance as new storage technology requires the commodity to be handled in water, where pathogens spores become more numerous and fruit contamination can occur. Combating these types of infections is difficult, since in many cases a pathogen is not very accessible. However, there is great potential for reaching a pathogen such as this with bacterial or yeast antagonists.

Fruit injuries, during and after harvest, present an excellent environment for disease development. Progress in horticulture necessitates mechanical harvest of some fruit, basically used for processing only,[16] mainly because of injury occurring during mechanical harvest. Even though injury may not play a major role in fruit harvested for processing, damage in the form of fresh wounds is the main site of entry for many important postharvest pathogens. The percentage of injured fruit could be lowered by careful handling; however, some open wound injuries such as stem pulls are unavoidable in many fruits.

C. Microbial Interaction of Fruit

The fruit surface is an area of constant interaction among saprophytic and parasitic organisms and their products. Fruit colonizing organisms can be categorized as beneficial, neutral, or deleterious. Similar to a phylloplane,[14,20,32,61] it will not be as necessary to return to equilibrium on the fruit surface as it is with biocontrol of soil-borne diseases.[5,19] In a dry climate where no irrigation is used the number of microorganisms is very small. This can be a disadvantage in field applied biocontrol, but in the case of postharvest application it may have no effect, or even could be helpful because of less competition from other organisms. The increasing occurrence of iatrogenic diseases in recent years shows the importance of a biotic balance. Rapid spread of previously unimportant diseases such as rhizopus rots on strawberry,[43] alternaria rot on citrus,[62] or mucor rots on pome fruits[65] after treatment with benomyl are good examples.

In biocontrol, resident antagonists may have certain advantages over casual antagonists as to the growth, surface colonization, and survivability.[12,19] However, under controlled conditions of greenhouse or storage houses, certain critical factors such as moisture, temperature, and even the nutritional base can be manipulated, allowing introduced foreign antagonists to become very successful biocontrol agents.[6,31,52]

Antagonists effective against some postharvest pathogens of apples[41,42] have been isolated from areas where the pathogen and susceptible host are present, but the disease does not occur.[5,19] In other instances antagonistic organisms to the apple scab (*Venturia inaequalis*) pathogen were isolated from apple leaves,[4,38] to citrus rots from rotten citrus,[63] or to strawberry blue mold (*Botrytis cinerea*) from rotten strawberries.[69]

D. Physical Interaction on Fruit

Surface characteristics such as waxes, hairs, and general porosity influence water retention on the fruit surface, a critical factor in maintaining microbial populations. Thickness and toughness of the surface may influence the susceptibility of fruit to injury, as well as the effectiveness of a pathogen in penetrating the host, causing an immediate or latent infection.

III. STRATEGIES IN FRUIT BIOCONTROL

Fruit disease control by microorganisms can be realized either directly, by applying organisms to fruit, or indirectly, by inoculum reduction. The degree to which fruit diseases should be controlled is generally higher than for other plant parts not utilized directly by the consumer. Thus, for biocontrol to be successful on fruit, complete elimination or inactivation of a pathogen should be achieved. Mere reduction in population of a pathogen does not suffice since it allows a pathogen to enter a host and cause some damage. Therefore, the question is whether biological control alone can protect fruit adequately from attack by pathogens. Examples of biocontrol of postharvest diseases of stone and pome fruits seem to prove that it can.[41,58] However, a certain strategy would have to be followed. Prevention of infection has the greatest chance of success. Inoculum reduction, e.g., in the case of *V. inaequalis* on apples[22] or *Botrytis cinerea* on grapes,[27] can be of help, but only when integrated with chemical control.

Fruit pathogens, many of which are unspecialized necrotrophs, depend on exogenous nutrients for germination and initiation of the pathogenic process.[10,11,13] Therefore, reduction in the nutrient base appears to be a good method to diminish chances of successful infection, particularly on wounds, which are the main point of entry for necrotrophic pathogens and are rich in nutrients and moisture. This nutrient depletion could well be accomplished by microorganisms which are very efficient colonizers, particularly at a wound site, and hopefully could withstand occasional dry conditions. Competition for nutrients and space, although sometimes considered controversial as to importance in nature,[20] may be very important in the development of many postharvest diseases, since necrotrophic parasites require a large food base for successful infection.

Antibiosis is another mode of action which could be very efficient in controlling diseases on fruit, mainly because the target area (fruit) is well defined and accessible, especially in postharvest situations. In addition, the energy supply needed to produce antibiotics may be available in fruit. Also, the buffering capacity of the target area is small compared to that of the soil ecosystem. Successful biocontrol of *Monilinia fructicola*, on peaches and other stone fruits, was achieved by Pusey and Wilson[56-58] using *Bacillus subtilis*. Antibiosis is probably the main mode of action in this case.

Stimulation of host (fruit) resistance is another possible mode of action for some antagonists. Observations with apples resistant to *Penicillium expansum*[42] suggests this, but conclusive data is still lacking.

Hyperparasitism may not be a good mode of action to protect against fruit diseases. With such organisms action is slower[5,45,60] than with antagonists having other modes of action; therefore, pathogen arrest may be too late in the case of fruit where an undamaged product is expected. Also, hyperparasites require earlier presence of the host (pathogen) for maintaining and increasing their population.

Compatibility of biocontrol agents with pesticides, e.g., residues from sprays, is important so that antagonists and their antibiotics are not inactivated, and also so that the antagonists do not destroy a pesticide. Synergistic activity is very desirable since it may allow a reduction in the amount of the chemical applied. It is desirable that biocontrol agent-pesticide compatibility allow the use of biocontrol agents before, after, or during chemical applications. Integrated control could involve chemical control in the field and biological control after harvest, where, in the controlled environment of storage rooms, it has the highest chance of success. In addition to the capability of controlling environmental factors, we do not have to be concerned with the effect of antagonists on plants in the postharvest environment as is the case of the biocontrol of soil-borne diseases in the field. Also, lack of damaging UV-spectrum light in storage may increase the number of potential antagonists.

A. Control of Small Fruit Rot

Most of the very limited work on small fruits microflora and biocontrol has been conducted on strawberries and is reviewed by Dennis.[26] Spoilage of small fruits such as strawberries, raspberries, blackberries, and blueberries is caused mainly by *Botrytis cinerea* and, in the case of the first three, also by *Mucor mucedo*. Microorganism composition[26] of freshly harvested strawberries, raspberries, and blackberries is somewhat similar; therefore, it is conceivable that some of the findings from work on strawberries will apply to the other genera in this group.

Grey-mold of strawberries, similar to dry-eye rot or grey-mold of apples, is incited by *B. cinerea,* a weak pathogen usually invading healthy strawberry fruits in the field through senile floral parts, and after harvest, through wounds on mature fruits.[55] Bhatt and Vaughan[8,9] isolated many fungi from strawberry flowers and fruit exhibiting antagonistic activity against *B. cinerea* in vitro on different agar media. In greenhouse experiments, three of those fungi, *Cladosporium herborum, Pullularia pullulans,* and the *Penicillium* sp., when tested as spore suspensions for control of the disease, gave reductions in rot of 42, 31, and 4%, respectively. However, when tested in the field no reduction in rot occurred, but an increase in yield was observed.

Tronsmo and Dennis[69] tested seven *Trichoderma* species, *Gliocladium virens,* and *Hypocrea semiorbitis* for antagonistic properties against *B. cinerea* and *M. mucedo,* another very important pathogen of strawberry which is generally known for its rapid spread through harvested fruit. Antagonists applied to strawberry plants as spore suspensions (10^7 spores/mℓ), in aqueous 0.01% tween 80 (containing 1% sucrose), from early flowering stage until 2 weeks before harvest showed (e.g., in the case of *T. viride*) good reduction in spoilage in the field and on stored strawberries. Recovery of viable *Trichoderma* from harvested fruit indicated the ability of the applied fungus to survive in the field. However, variations in tolerance to low temperature were observed, which probably accounts for variations in effectiveness among isolates. No reduction in *M. mucedo* rot occurred except from one *T. viride* isolate, while rot actually increased in two other cases. Similarly, in our research,[78] the application of different antagonists to strawberry plants in the field and to fruit after harvest, to reduce botrytis rot, also seems to increase mucor rot, but much variation between treatments has been observed. Are these instances of iatrogenic plant diseases caused by biocontrol agents? We should at least consider this possibility.

B. Control of Grey Mold on Grapes

Research on the biocontrol of grey-mold on grapes caused by *B. cinerea* was carried out in Bordeaux (France) for more than a decade.[27,28,29] Control efficiency of up to 70% (Abbott's formula) was obtained after repeated (four times) sprays of *Trichoderma harzianum* conidia (10^8 spores/mℓ) in vineyards, starting at bloom and finishing a few weeks before harvest.

It is known[17] that *B. cinerea* very effectively infects grapes through flowers, while infection of young berries is not observed. After infection, the fungus remains in the latent stage until it resumes growth during ripening. At the onset of maturation, spread of the disease occurs from mycelium growing from rotten to sound grapes and by conidial infection through the air. Vineyards abound with *B. cinerea* spores. The flower infection, similar to the earlier mentioned case with strawberries, emphasizes the importance of early application of the antagonist during bloom. Through colonization of senescent parts of flowers, *T. harzianum* disrupts the saprophytic phase of *B. cinerea*. Although the antagonist does not limit the number of infection sites, it definitely limits their development.[27] *Trichoderma harzianum* also reduced the number of conidia produced by sclerotia in the late spring.

Though not as effective as chemical treatment (80 to 95% efficiency), the antagonist could be used to control grey-mold as part of an integrated program permitting a substantial reduction in chemicals.

C. Control of Pineapple Fruit Diseases

Three pineapple (*Ananas comosum*) fruit diseases, interfruitlet corking (IFC), leathery pocket (LP), and fruitlet core rot (FCR), can be caused by *Penicillium funiculosum*.[59] Lim and Rohrbach[49] found that fungus strains isolated from pineapple (red-pigmented, reverse on PDA medium) and strains from sources other than diseased fruit were not pathogenic in contrast to nonpigmented (reverse on PDA medium) strains isolated from diseased fruit. If a spore suspension of the red-pigmented strain was applied as a spray to the heart of plants 1 and 4 weeks after chemical forcing, the severity of all 3 diseases was significantly reduced and the percentage of $2 \frac{1}{2}$ fruit size (too large to pass through a 12.7 cm diameter ring) increased greatly in comparison to treatment with pathogenic strains and the uninoculated check. Stimulation of phytoalexins in the pineapple plant and competition for nutrients were suggested as a possible explanation for disease reduction by these nonpathogenic strains. Although the disease was not totally eliminated the significant reduction indicates that application of this biocontrol agent could be made practical if more were known about such factors as the effect of inoculum level of the pathogen and the antagonist on disease development.

D. Control of Fungal Pathogens of Citrus Fruit

The idea of biological control of citrus fruit diseases dates back to the early fifties and the work of Gutter and Littauer[35] who observed instances of bacterial antagonism during isolations of pathogenic fungi from rotten citrus fruit. The bacterium was identified as *Bacillus subtilis*, an organism already known for antifungal properties. This antagonist, when tested in vitro on PDA medium against 13 major pathogens of citrus fruit, resulted, in most cases, in persistent inhibition zones. However, efforts to apply this principle to control citrus fruit pathogens was delayed another 30 years until extensive benomyl application caused a sudden increase in *Alternaria citri* rot.[15,62] This disease is apparently encouraged by pesticidal suppression of fungi antagonistic to the pathogen. This resulted in a search for new, effective pesticides and stimulated research on biological control of *A. citri* and other citrus pathogens.[47,63] Newly isolated *B. subtilis* strains exhibited good inhibition of *A. citri* and *Geotrichum candidum* in vitro, as well as in vivo, on Ellendale tangor of *Citrus reticulata* and Washington naval orange of *C. sinensis*. Antagonism in vivo was also observed against *Penicillium digitatum*, another important postharvest disease on citrus.

Bacterial spore powder, prepared by freeze drying of potato-dextrose-broth grown bacterial cells, maintained in vitro activity after resuspension in water or different media. The mode of action is probably through antibiosis, but no proof of this has been presented.

E. Control of Stone Fruit Rots

Brown rot of stone fruits incited by *Monilinia fructicola* is a major postharvest disease of the *Prunus* spp. Pusey and Wilson[56-58,77] isolated the bacterium *Bacillus subtilis* (B-3), which was strongly inhibitory to *M. fructicola* in vitro on PDA or nutrient-yeast-dextrose agar (NYDA) media, and in vivo on peaches, nectarines, apricots and plums (Figure 1). They applied the antagonist (B-3) to fruit as a water suspension or as a mixture with nutrient-yeast-dextrose broth (NYDB) or waxes. The treatment was more effective on peaches and apricots than on nectarines and plums. The authors speculated that these results may be related to basic differences in the fruit surfaces effecting bacterial adherence. The antagonist concentration of 10^8 colony forming units (CFU) /mℓ was sufficient to inhibit brown rot development on all of the above mentioned fruit inoculated with 10^5 spores/mℓ of *M.*

FIGURE 1.   Postharvest reduction of brown rot of peaches by antagonistic bacteria *Bacillus subtilis* (B-3). Wounded peaches 3 days after treatment with (left) water or (right) *Bacillus subtilis*, then inoculation with *Monilinia fructicola*.

fructicola. The antagonist was active from 5 to 30°C when tested on peach fruit. The culture filtrate was as effective as the antagonist cell suspension. Autoclaving only slightly decreased activity of the filtrate, but totally eliminated the effectiveness of the bacterial cells, indicating that the B-3 isolate produced a heat-stable antifungal compound. Once lesions on fruit were initiated they enlarged at a constant rate indicating that antifungal activity is limited to spore germination inhibition with little effect on subsequent growth stages. The bacterium is also compatible to dicloran, which is commonly used to control rhizopus rot, an important disease not controlled by strain B-3. Currently, a pilot test is in progress to investigate the feasibility of using this system on a commercial scale.[79]

A recently isolated strain of the bacterium *Enterobacter cloacae* (D-1) exhibited antagonistic activity against *Rhizopus stolonifer*, a pathogen of stone fruits.[80] Antagonism was observed in vitro on peaches, although total control was not obtained. Filtrates obtained from the bacterial growth on NYDB medium had no antibiotic activity. This, in addition to accumulation of the antagonist cells around fungal structures, suggests a mode of action different from *B. subtilis* antibiosis. These observations seem to agree with other findings[37] which demonstrated that *E. cloacae* cells contain agglutinins capable of binding to various compounds of the fungal (*Pythium* spp.) cell wall, subsequently causing their lysis. It would be interesting to study the compatibility of these two bacterial antagonists and to further investigate their effectiveness in combating their respective diseases.

F. Control of Dry Eye Rot of Apples

Dry eye rot disease of apples is caused by *Botrytis cinerea*. The infection begins in the orchard during bloom when sepals become infected and the fungus expands into fruit tissue generally no more than 3 cm deep. The rot then dries and the pathogen is retained behind a lignin barrier.[71,72] After the fruit matures, rot can expand further and the fungus can infect tissue from adjacent fruit. Tronsmo and Raa[73] observed antagonistic action of *Trichoderma pseudokeningii* and *T. harzianum* against *B. cinerea*. By repeated spore application of *T. pseudokeningii* (1×10^6/mℓ) alone, or mixed with *B. cinerea* (5×10^4 spores/mℓ), they obtained a reduction in the percentage of dry eye rot when compared to *B. cinerea* treatment alone.

However, they did not obtain significant reduction in naturally occurring infection in the field. This failure was attributed to a lack of low temperature tolerance in the antagonist. In a later study[70,74] a *T. harzianum* antagonistic isolate selected for low temperature tolerance applied with a food base gave good control of the disease in the field.

G. Control of Apple Scab

Andrews et al.[4] isolated, from apple leaves, fungal antagonists to the apple scab pathogen, *Venturia inaequalis*. After three in vitro tests (i.e., growth on nutrient agar, germination, and measurements of germ tube lengths on agarose coated slides) and three in vivo tests (i.e., lesion size, symptom development, and conidia production), the highest ranked antagonist was *Chaetomium globosum*. Other antagonists selected from 50 organisms included *T. viride*, *Microsphaeropsis olivacea*, *Aureobasidium pullulans*, the *Cryptococcus* sp., an actinomycete, and the *Flavobacterium* sp. In later studies[38] treatment with the basidiomyceteous antagonist *Athelia bombacina* resulted in control superior to that obtained with *C. globosum*. The mechanisms of action in this case were nutrient competition and antibiosis.[21,38] This is an example of biocontrol by reduction of pathogen inoculum level.

H. Control of Blue Mold of Apples

Control of blue–mold caused by *Penicillium expansum* is one of the most promising biocontrol systems on fruit.[41,42] The antagonistic bacterium, L-22-64, identified by fatty acid profile as *Pseudomonas syringae* pv. *lachrymans,* was isolated from apple. Though it may be considered as a casual organism to apple, it is also an established resident on other fruit (cucumber). The natural fruit habitat of the antagonist could be a distinct advantage in using it as a biocontrol agent, since the organism is already adapted to a specific fruit niche.

No significant reduction in disease was observed when wounded, ripe Golden Delicious apples were dipped in bacterial suspension for only 30 sec. However, when dipping time was increased to 5 min significant protection occurred. This may be a direct result of the bacterial attachment to the fruit. Nonselective attachment of this bacterium to different plant surfaces was described by others.[18,36,48] Apples treated at optimum stage of maturity for storage responded even better to the treatment, resulting in total protection from the disease (Table 1). The concentration of the bacterium necessary for apple protection depended on the pathogen spore concentration used. An antagonist concentration of 8×10^7 CFU/mℓ was sufficient to prevent lesion development in a challenge with a pathogen inoculum spore concentration of 1×10^5 spores/mℓ. When the pathogen inoculum level was raised to 1×10^7 spores/mℓ, the lowest totally effective concentration of the antagonist was 4×10^8 CFU/mℓ. In the case of yeast antagonist F-43-31, which was isolated from apple fruit, concentrations for effective control were in both cases 1×10^8 CFU/mℓ. The pathogen spore concentrations in warehouse drench tanks are estimated to be somewhere between 10^3 and 10^4 spores/mℓ.[65]

Table 1
LESION DEVELOPMENT (MM) ON GOLDEN DELICIOUS APPLES AFTER PROTECTION WITH DIFFERENT CONCENTRATIONS OF BACTERIAL STRAIN L-22-64, INOCULATION WITH CONCENTRATIONS OF VARIOUS SPORE SUSPENSIONS OF *P. EXPANSUM* AND REINOCULATION AFTER 10 DAYS

Bacterial concentration (CFU/mℓ)	Pathogen spore concentrations				
	1×10^7	1×10^6	1×10^5	1×10^4	1×10^3
8×10^7	62.7	26.7	0.0	0.0	0.0
9×10^7	46.3	0.0	0.0	0.0	3.0
2×10^8	42.7	0.0	0.0	0.0	0.0
4×10^8	0.0	0.0	0.0	0.0	0.0
6×10^8	0.0	0.0	0.0	0.0	0.0
8×10^8	0.0	5.3	0.0	0.0	0.0
O-CK-P[w]	89.3	74.3	82.3	82.0	69.3
O-CK-W-P[x]	0.0	0.0	0.0	0.0	0.0

Note: [w] *Penicillium expansum* check.
[x] Water check inoculated with *P. expansum* only during reinoculation procedure.

Data adapted from Janisiewicz, W. J., *Phytopathology,* 77, 481, 1986. With permission.

IV. PERSPECTIVES FOR THE FUTURE

How long will it be before some of the findings which have been presented can be used on a commercial scale? Which direction should be taken in the near future? Undoubtedly, among the many concepts in biological control, fruit protection with biological agents (particularly those applied after harvest) is getting deservedly increasing attention. The instances of biocontrol cited here demonstrate a great diversity of organisms capable of accomplishing this task. Pseudomonads, spore forming Bacilli, yeast, and filamentous fungi — each of these groups has a representative capable of providing biological control of fruit diseases. Just as there is diversity among organisms, there is also diversity of mechanisms by which they operate. Although in the majority of cases these mechanisms have not been satisfactorily elucidated, they probably involve antibiosis, nutrient competition, stimulation of host defense, predation, and parasitism. It is very probable that, in many cases, more than one mechanism operates.

What would be a good biocontrol agent for fruit? Studies of the most successful example, *Agrobacterium tumefaciens* strain K 84, used in protection of trees in nurseries against crown gall, indicate that the critical period of protection is from the initiation to the healing of a wound.[44,60,61] Since many important fruit diseases (especially from the postharvest group) are wound pathogens, a biocontrol system allowing wounds to heal prior to pathogen invasion seems very desirable. An example of such a system is blue–mold of apples controlled by *P. syringae* pv. *lachrymans* (a cucumber pathogen) where, after protection of the fresh wound for a period of time, the lesions become resistant to pathogen invasion.[42]

Are fruit pathogens potentially successful candidates for biocontrol? Fruit pathogens are ecologically fit to the specific niche created by fruit, and are therefore good candidates to control fruit diseases. As suggested by Schroth et al.,[61] "the most effective biocontrol agent may be an avirulent, antagonistic strain derived from or related to the pathogen." From this perspective, using a casual fruit pathogen (as in the case of blue mold control on apples) or a resident, but avirulent pathogen, as presented in the case of biocontrol of IFC, LP, and FCR diseases of pineapple, seems to be most promising. However, using pathogens as biocontrol agents raises questions as to the safety of the method. The possibility of using "proper" nonpathogenic mutants (i.e., deletion mutants) offers strong support for this method.[50]

Another question recently raised concerns the possibility of the natural development of pathogen strains resistant to biocontrol agents. Such an example was well documented in the case of *A. tumefaciens* strain K 84 in soils of different parts of the U.S.[1,60] and other countries.[54] We should be aware of this possibility, although in certain instances, i.e., postharvest fruit treatment, such an occurrence is more remote than in field application. This is similar to the benlate situation in the apple and pear industry of Washington State. Restriction to only postharvest use prevented the development of resistant strains of pathogens for a long time.[65] Using a mixture of antagonistic organisms with different modes of action, or a single organism with a few different antagonistic mechanisms, could lower the possibility of such resistance developing.

Another problem which can develop with biocontrol agents is the appearance of "iatrogenic" diseases. Little has been done in this area, but some observations after treatment with biocontrol agents suggest that this may be a potential problem.

What are the major obstacles hindering full development of fruit biocontrol? The answer seems to lie in problems common to most biocontrol approaches — a lack of consistent, positive results in field trials and in the economics of the approach.[60,61] Some research remains incomplete after only partial success in the first few trials. Perhaps work should have continued toward a better understanding of the ecology and physiology of antagonistic organisms and their interactions with pathogens. For instance, in the case of apple dry eye rot control by *T. harzianum*,[68,70,74] simply obtaining low temperature tolerant isolates was sufficient to drastically alter the situation, making field application of the antagonist very successful.

Another obstacle is the lack of knowledge of factors which favor antagonists on fruits, as are known for soil systems. Studies of microflora changes on apples[24,46,76] and on small fruits[26,64,75] during fruit maturation indicate that certain nutrients favor particular organisms. Knowing the nutritional requirements of an antagonist would be important in manipulating its population; also, knowing pH and nutrient composition on the fruit surface would help in the selection of proper antagonists.

Drought resistance of microorganisms is a major factor affecting fruit biocontrol development. Yeasts, unlike many bacteria, can colonize the surface for long periods under dry conditions. Large amounts of extra-cellular polysaccharides contribute to their survivability. By having a favorable surface-to-volume ratio (in contrast to mycelial fungi) yeasts can utilize available nutrients, increase in number, and colonize the surface (particularly at the wound site) very rapidly. Yeasts are also organisms on which pesticide impact is minimal.[2,3,40] Furthermore their genetics are relatively well-known and thus they can be manipulated through genetic engineering. This, together with antagonistic activity and the ability to fill a niche after "negative alternation" (niche created after destroying nontarget microflora by pesticides), is another reason to consider yeasts as very good candidates for biocontrol of fruit diseases. Finally, public acceptance of yeasts as organisms applied directly to consumed commodities assures them a better chance than other organisms of more questionable ingestive properties.

The economic aspect of biocontrol has taken a different twist in recent years. As already mentioned, new effective chemicals are increasingly more difficult to develop. Therefore, with some pathogens (i.e., *Mucor* spp., *Alternaria* spp.) where chemical control is lacking, the use of biocontrol agents may become the best alternative. The concept of economics, in the case of fruit, particularly those already harvested, acquires a slightly different meaning. Harvested fruit represents a high-value investment of an entire growing season. Application of disease protective methods too expensive in other crops may prove to be economically feasible for protection of harvested fruit.

Considerable pioneering in the field of fruit biocontrol has already taken place, resulting in many promising avenues for further exploration. All things considered, it would seem that the time is ripe to move zealously into the development of practical systems of biocontrol of diseases of fruit.

REFERENCES

1. **Alconero, R.,** Crown gall of peaches from Maryland, South Carolina, and Tennessee and problems with biological control. *Plant Dis.,* 64, 835, 1980.
2. **Andrews, J. H.,** Effects of pesticides on non-target micro-organisms on leaves, in *Microbial Ecology of the Phylloplane,* Blakeman, J. P., Ed., Academic Press, New York, 1981, 284.
3. **Andrews, J. H. and Kenerley, C. M.,** The effects of a pesticide program on non-target epiphytic microbial populations of apple leaves, *Can. J. Microbiol.,* 24, 1058, 1978.
4. **Andrews, J. H., Berbee, F. M., and Nordheim, E. V.,** Microbial antagonism to the imperfect stage of the apple scab pathogen, *Venturia inaequalis, Phytopathology,* 73, 228, 1983.
5. **Baker, K. F. and Cook, R. J.,** *Biological Control of Plant Pathogens,* W. H. Freeman & Co., San Francisco, 1974, 433.
6. **Bashi, E. and Fokkema, N. J.,** Environmental factors limiting growth of *Sporoblolomyces roseus,* an antagonist of *Cochliobolus sativus,* on wheat leaves, *Trans. Br. Mycol. Soc.,* 68, 17, 1977.
7. **Battenfield, S. L., Ed.,** Proc. Natl. Interdisciplinary Biological Control Conf., Las Vegas, Nevada, February 15 to 17, 1983, 4.
8. **Bhatt, D. D. and Vaughan, E. K.,** Preliminary investigations on biological control of gray mold (*Botrytis cinerea*) of strawberries, *Plant Dis. Rep.,* 46, 342, 1962.
9. **Bhatt, D. D. and Vaughan, E. K.,** Interrelationships among fungi associated with strawberries in Oregon, *Phytopathology,* 53, 217, 1963.
10. **Blakeman, J. P.,** Effect of plant age on inhibition of *Botrytis cinerea* spores by bacteria on beetroot leaves, *Physiol. Plant Pathol.,* 2, 143, 1972.
11. **Blakeman, J. P.,** The chemical environment of leaf surfaces with special reference to spore germination of pathogenic fungi, *Pestic. Sci.,* 4, 575, 1973.
12. **Blakeman, J. P.,** Ecological succession of leaf surface microorganisms in relation to biological control, in *Biological Control on the Phylloplane,* Windles, C. E. and Lindow, S. E., Eds., American Phytopathological Society, St. Paul, Minn., 1985, 6.
13. **Blakeman, J. P. and Fraser, A. K.,** Inhibition of *Botrytis cinerea* spores by bacteria on the surface of chrysanthemum leaves, *Physiol. Plant Pathol.,* 1, 45, 1971.
14. **Blakeman, J. P. and Fokkema, N. J.,** Potential for biological control of plant diseases on the phylloplane, *Annu. Rev. Phytopathol.,* 20, 167, 1982.
15. **Brown, G. E. and McCornack, A. A.,** Decay caused by *Alternaria citri* in Florida citrus fruit, *Plant Dis. Rep.,* 56, 909, 1972.
16. **Brown, G. K.,** Harvest Mechanization Status for Horticultural Crops, American Society of Agricultural Engineers, Paper No. 80-1532, St. Joseph, Mich., 1980.
17. **Bulit, J. and LaFon, R.,** Biologie due *Botrytis cinerea* Pers. et le development de la pourriture grise de la vigne, *Rev. Zool. Agric. Pathol. Veg.,* 71, 1, 1972.
18. **Chet, I., Zilberstein, Y., and Henis, Y.,** Chemo-taxis of *Pseudomonas lachrymans* to plant extracts and to water droplets collected from the leaf surfaces of resistant and susceptible plants, *Physiol. Plant Pathol.,* 3, 473, 1973.
19. **Cook, R. J. and Baker, K. F.,** *The Nature and Practice of Biological Control of Plant Pathogens,* American Phytopathological Society, St. Paul, Minn., 1983, 539.
20. **Cullen, D. and Andrews, J. H.,** Epiphtyic microbes as biological control agents, in *Plant-Microbe Interactions: Molecular and Genetic Perspectives,* Kosuge, T. and Nester, E. W., Eds., Macmillian, New York, 1984, 381.

21. **Cullen, D. and Andrews, J. H.,** Evidence for the role of antibiosis in the antagonism of *Chaetomium globosum* to the apple scab pathogen, *Venturia inaequalis, Can. J. Bot.,* 62, 1819, 1984.

22. **Cullen, D., Berbee, F. M., and Andrews, J. H.,** *Chaetomium globosum* antagonizes the apple scab pathogen, *Venturia inaequalis,* under field conditions, *Can. J. Bot.,* 62, 1814, 1984.

23. **Davenport, R. R.,** Ecological concepts in studies of micro-organisms on aerial plant surfaces in *Microbiology of Aerial Plant Surfaces,* Dickinson, C. H. and Preece, T. F., Eds., Academic Press, New York, 1976, 199.

24. **Davenport, R. R.,** Distribution of yeasts and yeast-like organisms from aerial surfaces of developing apples and grapes, in *Microbiology of Aerial Plant Surfaces,* Dickinson, C. H. and Preece, T. F., Eds., Academic Press, New York, 1976, 325.

25. **Dennis, C., Ed.,** *Post-Harvest Pathology of Fruits and Vegetables,* Academic Press, New York, 1983, 266.

26. **Dennis, C.,** The microflora of the surface of soft fruits, in *Microbiology of Aerial Plant Surfaces,* Dickinson, C. H. and Preece, T. F., Eds., Academic Press, New York, 1976, 419.

27. **Dubos, B.,** Biocontrol of *Botrytis cinerea* on grapevines by antagonistic strain of *Trichoderma harzianum,* in *Current Perspectives in Microbial Ecology,* Klub, M. J. and Reedy, C. A., Eds., American Society for Microbiology, Washington, D.C., 1984, 370.

28. **Dubos, B. and Bulit, J.,** Filamentous fungi as biocontrol agents on aerial plant surfaces, in *Microbiol Ecology of the Phylloplane,* Blakeman, J. P., Ed., Academic Press, New York, 1981, 353.

29. **Dubos, B., Jailloux, F., and Bulit, J.,** Employing antagonistic properties of *Trichoderma* against *Botrytis cinerea* in the protection of vineyards against grey-mold, *Phytoparasitica,* 10, 134, 1982.

30. **Eckert, J. W.,** Postharvest diseases of fresh fruits and vegetables — etiology and control, in *Postharvest Biology and Handling of Fruit and Vegetables,* Haard, N. F. and Salunkhe, D. K., Eds., CRC Press, Boca Raton, Fla., 1975, 81.

31. **Fokkema, N. J., Houter, J. G. den, Kosterman, Y. J. C., and Nelis, A. L.,** Manipulation of yeasts on field-grown wheat leaves and their antagonistic effect on *Cochliobolus sativus* and *Septoria nodorum, Trans. Br. Mycol. Soc.,* 72, 19, 1979.

32. **Fokkema, N. J., Van der Laar, J. A. J., Nelis-Blomberg, A. L., and Schippers, B.,** The buffering capacity of the natural mycoflora of rye leaves to infection by *Cochliobolus sativus,* and its susceptibility to benomyl, *Neth. J. Plant Pathol.,* 81, 176, 1975.

33. **Fokkema, N. J. and Van der Meulien, F.,** Antagonism of yeast-like phyllosphere fungi against *Septoria nodorum* on wheat leaves, *Neth. J. Plant Pathol.,* 82, 13, 1976.

34. **Griffiths, E.,** Iatrogenic plant diseases, *Annu. Rev. Phytopathol.,* 19, 69, 1981.

35. **Gutter, Y. and Littauer, F.,** Antagonistic action of *Bacillus subtilis* against citrus fruit pathogens, *Bull. Res. Counc. Isr.,* 3, 192, 1953.

36. **Haas, J. H. and Rotem, J.,** *Pseudomonas lachrymans* adsorption, survival, and infectivity following precision inoculation of leaves, *Phytopathology,* 66, 992, 1976.

37. **Hadar, Y., Harman, G. E., Taylor, A. G., and Norton, J. M.,** Effects of pregermination of pea and cucumber seeds and of seed treatment with *Enterobacter cloacae* on rots caused by *Pythium* spp., *Phytopathology,* 73, 1322, 1983.

38. **Heye, C. C. and Andrews, J. H.,** Antagonism of *Athelia bombacina* and *Chaetomium globosum* to the apple scab pathogen, *Venturia inaequalis, Phytopathology,* 73, 650, 1983.

39. **Hislop, E. C.,** Some effects of fungicides and other agrochemicals on the microbiology of the aerial surfaces of plants, in *Microbiology of Aerial Plant Surfaces,* Dickinson, C. H. and Preece, T. F., Eds., Academic Press, New York, 1976, 41.

40. **Hislop, E. C. and Cox, T. W.,** Effects of captan on the non-parasitic microflora of apple leaves, *Trans. Br. Mycol. Soc.,* 52, 223, 1969.

41. **Janisiewicz, W. J.,** Biological control of post-harvest diseases of pome fruits, *Phytopathology,* 75 (Abstr.), 1301, 1985.

42. **Janisiewicz, W. J.,** Postharvest biological control of blue-mold on apples, *Phytopathology,* 77, 481, 1986.

43. **Jordan, V. W. L.,** The effects of prophylactic spray programes on the control of pre- and post-harvest diseases of strawberry, *Plant Pathol.,* 22, 67, 1973.

44. **Kerr, A.,** Biological control of crown gall through production of agrocin 84, *Plant Dis.,* 64, 25, 1980.

45. **Kranz, J.,** Hyperparasitism of biotrophic fungi, in *Microbial Ecology of the Phylloplane,* Blakeman, J. P., Ed., Academic Press, New York, 1981, 327.

46. **Last, F. T. and Price, D.,** Yeasts associated with living plants and their environs, *Yeast,* 1, 183, 1970.

47. **Laville, E. Y., Harding, P. R., Dagan, Y., Rhat, M., Kraght, A. J., and Rippon, E. L.,** Studies on imazalil as potential treatment for control of citrus fruit decay, *Proc. Int. Soc. Citricult.,* 1, 269, 1977.

48. **Leben, C. and Whitmoyer, R. E.,** Adherence of bacteria to leaves, *Can. J. Microbiol.,* 25, 896, 1979.

49. **Lim, T. and Rohrbach, K. G.,** Role of *Penicillium funiculosum* strains in the development of pineapple fruit diseases, *Phytopathology,* 70, 663, 1980.

50. **Lindemann, J.,** Genetic manipulation of microorganisms for biological control, in *Biological Control on*

the Phylloplane, Windels, C. E. and Lindow, S. E., Eds., American Phytopathological Society, St. Paul, Minn., 1985, 169.

51. **Lindow, S. E.,** Integrated control and role of antibiosis in biological control of fireblight and frost injury, in *Biological Control on the Phylloplane,* Windels, C. E. and Lindow, S. E., Eds., American Phytopathological Society, St. Paul, Minn., 1985, 169.

52. **Morris, C. E. and Rouse, D. I.,** Role of nutrients in regulating epiphytic bacterial populations, in *Biological Control on the Phylloplane,* Windels, C. E. and Lindow, S. E., Eds., American Phytopathological Society, St. Paul, Minn., 1985, 169.

53. **Napoli, C. and Staskawicz, B.,** Molecular genetics of biological control agents of plant pathogens: status and prospects, in *Biological Control in Agricultural IPM Systems,* Hoy, M. A. and Herzog, D. C., Academic Press, New York, 1985, 455.

54. **Panagopoulos, C. G., Psallidas, P. G., and Alvizatos, A. S.,** Evidence of a breakdown in the effectiveness of biological control of crown gall, in *Soil-borne Pathogens,* Schippers, B. and Gams, W., Eds., Academic Press, New York, 1979.

55. **Powelson, R. L.,** Initiation of strawberry fruit rot caused by *Botrytis cinerea, Phytopathology,* 50, 491, 1960.

56. **Pusey, P. L. and Wilson, C. L.,** Effect of bacterial antagonists on fungal rots of decidious fruit, *Phytopathology,* 72 (Abstr.), 710, 1982.

57. **Pusey, P. L. and Wilson, C. L.,** Control of brown rot with a bacillus bacterium, *Phytopathology,* 73 (Abstr.), 823, 1983.

58. **Pusey, P. L. and Wilson, C. L.,** Postharvest biological control of stone fruit brown rot by *Bacillus subtilis, Plant Dis.,* 68, 753, 1984.

59. **Rohrbach, K. G. and Pfeiffer, J. B.,** Field induction and etiology of pineapple interfruitlet corking, leathery pocket and fruitlet core rot with *Penicillium funiculosum, Phytopathology,* 66, 392, 1976.

60. **Schroth, M. N. and Hancock, J. G.,** Selected topics in biological control, *Annu. Rev. Microbiol.,* 35, 453, 1981.

61. **Schroth, M. N., Loper, J. E., and Hildebrand, D. C.,** Bacteria as biocontrol agents of plant disease, in *Current Perspectives in Microbial Ecology,* Klug, M. J. and Reedy, C. A., Eds., American Society for Microbiology, Washington, D.C., 1984, 710.

62. **Singh, V.,** Control of alternaria rot in citrus fruit, *Australas. Plant Pathol.,* 9, 12, 1980.

63. **Singh, V. and Deverall, B. J.,** *Bacillus subtilis* as a control agent against fungal pathogens of citrus fruit, *Trans. Br. Mycol. Soc.,* 83, 487, 1984.

64. **Singh, J. P. and Kainsa, R. L.,** Microbial flora of grapes in relation to storage and spoilage, *Indian Phytopathol.,* 36, 72, 1983.

65. **Spotts, R. A. and Cervantes, L. A.,** Populations, pathogenicity, and benomyl resistance of *Botrytis* spp., *Penicillium* spp., and *Mucor piriformis* in packinghouses, *Plant Dis.,* 70, 106, 1986.

66. **Spurr, H. W., Jr.,** Experiments on foliar disease control using bacterial antagonists, in *Microbial Ecology of the Phylloplane,* Blakeman, J. P., Ed., Academic Press, New York, 1981, 369.

67. **Swinburne, T. R.,** Quiescent infections in post-harvest diseases, in *Post-Harvest Pathology of Fruits and Vegetables,* Dennis, C., Ed., Academic Press, New York, 1983, 1.

68. **Tronsmo, A.,** *Trichoderma harzianum* used as a biocontrol agent against *Botrytis cinerea* on apple, *Colloq. I'INRA,* 18, 109, 1983.

69. **Tronsmo, A. and Dennis, C.,** The use of *Trichoderma* species to control strawberry fruit rots, *Neth. J. Plant Pathol.,* 83, 449, 1977.

70. **Tronsmo, A. and Dennis, C.,** Effect of temperature on antagonistic properties of *Trichoderma* species, *Trans. Br. Mycol. Soc.,* 71, 469, 1978.

71. **Tronsmo, A. and Raa, J.,** Morphological/cytological description of dry eye rot in apple fruit caused by *Botrytis cinerea* Pers., *Acta Agricul. Scan.,* 28, 218, 1978.

72. **Tronsmo, A. and Raa, J.,** Life cycle of the dry eye rot pathogen *Botrytis cinerea* Pers. on apple, *Phytopathol. Z.,* 89, 203, 1977.

73. **Tronsmo, A. and Raa, J.,** Antagonistic action of *Trichoderma pseudokoningii* against the apple pathogen *Botrytis cinerea, Phytopathol. Z.,* 89, 216, 1977.

74. **Tronsmo, A. and Ystaas, J.,** Biological control of *Botrytis cinerea* on apple, *Plant Dis.,* 64, 1009, 1980.

75. **Wadia, K. D., Manoharachary, C., and Janaki, C. H.,** Fruit surface mycoflora of *Vitis vinifera* L. and *Capsicum annuum* L. in relation to their fruit rot diseases, *Proc. Indian Nat. Sci. Acad.* B49, 371, 1983.

76. **Williams, A. J., Wallace, R. H., and Clark, D. S.,** Changes in the yeast population on Quebec apples during ripening, *Can. J. Microbiol.,* 2, 645, 1956.

77. **Wilson, C. L. and Pusey, P. L.,** Potential for biological control of postharvest plant diseases, *Plant Dis.,* 69, 375, 1985.

78. **Janisiewicz, W. J.,** unpublished data.

79. **Wilson, C. L.,** personal communication.

80. **Wilson, C. L.,** personal communication.

Chapter 12

ANTAGONISM AND BIOLOGICAL CONTROL

J. Singh and J. L. Faull

TABLE OF CONTENTS

I. INTRODUCTION

In natural environments microorganisms interact as a consequence of their growth and development. Odum[45] demonstrated that the population of two species may interact in several different ways, and he designated these interaction categories as neutralism, competition, amensalism, parasitism, mutualism, protocooperation, commensalism, and predation. The microbial antagonism that is seen in biological control of plant pathogens is broadly based on the categories of competition (for nutrients and space), parasitism (which may be by the production of volatile or nonvolatile antibiotics), and hyperparasitism. All of these mechanisms may operate together or independently, and their activities can result in the suppression of microbial plant disease. It is the intention of the reviewers to describe these categories of interaction in turn, and to relate these mechanisms to examples of successful biological control in the field.

II. MODES OF ANTAGONISM

A. Competition
Competition occurs when there is demand by two or more microorganisms for the same resource in excess of the immediate supply. The term can be used broadly to denote factors favoring one species over another or in a narrower sense it can mean the active demand in excess of supply of material or space on the part of two or more organisms.[47,65]

Clark[14] defined competition as the injurious effect of one organism on another because of the utilization or removal of some resource of the environment. These resources can include nutrients, oxygen, and space.

Intraspecific and interspecific competition among microorganisms is one of the most important factors determining the density of population of an organism in nature. Species that prefer the same ecological niche usually avoid competition and occupy different geographical areas or different habitats in the same area.[45] Space and oxygen are probably important variables that can change in the rhizosphere or rhizoplane to the detriment of pathogen establishment, in that the occupation of a site by microorganisms must diminish the limited supply of nutrients in either environment. According to Baker[4] carbon, nitrogen, and vitamins are all important in this respect because they determine the growth and infection of soil-borne plant pathogens in competition with other organisms.

B. Hyperparasitism
The terms hyperparasitism, mycoparasitism, direct parasitism, and interfungus parasitism are used in reference to the phenomenon of one fungus parasitizing another.[8] Hyperparasitism covers a multitude of different interactions including minor or major morphological disturbances, the overgrowth of hyphae of one fungus by another, penetration and direct parasitism by the production of haustoria, and the lysis of one hyphae by another.[5,8,20] A form of hyperparasitism termed hyphal interference has been described by Ikediugwu and Webster[33] in a range of coprophilous fungi on rabbit dung and in the interaction between *Peniophora gigantea* and *Heterobasidium annosum* on tree stumps.[33,34] The interaction is characterized by the rapid cessation of growth followed by hyphal death when *H. annosum* hyphae contact *P. gigantea* hyphae.

Hyperparasitic interactions are of biotrophic and necrotrophic type, dependent upon the mode of parasitism, and on its effect on the host. Biotrophic parasites secure nutrients from the living cells of the host via haustoria. Necrotrophic parasites acquire nutrients from dead host cells, usually killed by the parasite before it invades, by toxins or extracellular enzymes. Coiling and anastomosis of one mycelium around another are indicative of some hyperparasitic interactions, and this can be a characteristic of biotrophic or necrotrophic associations. The resting structures of fungi may also be attacked by hyperparasitic fungi. The degree of parasitism is markedly affected by many factors, such as changes in the C:N ratio, changes in temperature, light, pH, and other nutrients.

C. Antibiosis

Antibiosis occurs when the production of antibiotics or toxic metabolites by microorganisms has a direct effect on another microorganism. Such compounds may be volatile or nonvolatile and their importance in the rhizosphere and on the phyllosphere is yet to be conclusively proven. The widespread production of antibiotics was reported by many authors,[10] but the ecological importance of these compounds was doubted by Park.[48] Many fungi have been shown to be capable of producing volatile[19,24,32] and nonvolatile antibiotics.[18,26] According to Baker[4] antibiotic production requires that the organism producing the substances has adequate food resources, particularly organic substrates like straw and seed coats in soil, and pollen on the phylloplane. The detection and extraction of antibiotics from the soil, the rhizosphere, or the phylloplane is very difficult, because they are absorbed on to the clay colloids and humus particles in the soil, and they are lost to the atmosphere on the phylloplane.[12] In all environments they are likely to be degraded by other microorganisms that are not adversely affected by them. According to Bruehl et al.[13] antibiotic production in microhabitats may be ecologically significant as these antibiotic producing microorganisms occupy small substrate niches in defiance of all intruders. Several fungi that inhabit the phylloplane of various plants have been shown to be capable of producing antibiotics. These include *Aureobasidium*, *Alternaria*, *Botrytis*, *Sporobolomyces*, and the *Helminthosporium* spp.[59]

III. PRACTICAL APPROACHES FOR IDENTIFICATION OF MODES OF ANTAGONISM

The potential antagonistic ability of any given microorganism can be assessed by a detailed screening program. This involves many different techniques to detect the varying classes of antagonism that can occur. A dual culture system on agar plates of the target and the potential antagonist can detect several different classes of antagonism, including antibiosis, competition, and hyperparasitism.

Competition on plates is easily seen by observing and recording the percentage area of the plate that is covered by each of the interacting microorganisms over a number of days. The faster growing of the two microorganisms will occupy the greater proportion of the plate initially, and if it is the better competitor it will come to dominate the plate over a few days. If it also possesses other features of antagonism towards the target fungus such as antibiosis or hyperparasitism, then it will very rapidly eradicate the slower growing organism.

Hyperparasitism between two microorganisms is best observed by microscopy, preferably by scanning and transmission electron microscopy. An easy method of preparing the material for examination by this method involves their coplating on a sterile cellophane membrane overlaid on media on an agar plate. The cultures can then be removed from the surface of the agar on the cellophane with minimum damage, and can be prepared for electron microscopy. The types of interaction seen in the scanning electron microscope of such material will include the coiling of one mycelia around another, often with obvious morphological disturbances in the target mycelium (Figure 1), and occasionally the penetration of the target mycelium can also be seen, although this must be confirmed by transmission microscopy.

Antibiosis can often be seen as a zone of inhibition that occurs between two microorganisms when they are grown together on agar plates. Such zones can be the result of volatile or

FIGURE 1. (A) *Thrichoderma harzianum* (TH) hyphae coiling around *Rhizoctonia solani* hyphae (RS) × 400 (Courtesy of R. Scarselletti). (B) *Trichoderma harzianum* (TH) hyphae coiling and penetrating the hyphae of *Rhizoctonia solani*. The TH hyphae are forming chlamydospores (C) × 350 (Courtesy of R. Scarselletti). (C) *Trichoderma harzianum* (TH) hyphae sporulating heavily after the collapse of *Rhizoctonia solani* (RS) hyphae. × 600 (Courtesy of R. Scarselletti).

nonvolatile antibiotic production by one of the two organisms. The two effects can be distinguished from each other by growing the organisms on split plates where only volatile antibiotics will be able to have their effects on the radial growth rate of the target fungus (Figure 2A). Nonvolatile antibiotic production can be detected by placing a small quantity

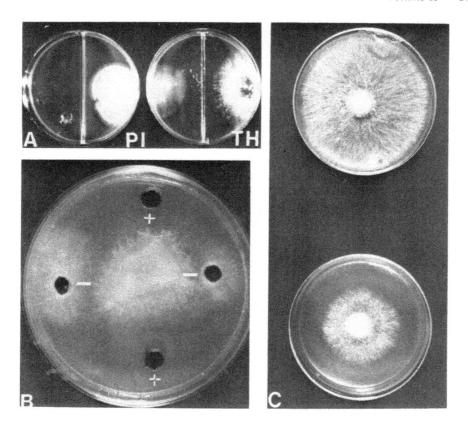

FIGURE 2.   (A) Split plate technique for detecting the effects of volatile antibiotics. Left *Phytophthora infestans* (PI) Right *Phytophthora infestans* + *Trichoderma harzianum TH* (B) Agar plate well technique for detecting nonvolatile antibiotics. (−) wells no filtrate, (+) wells filtrate added. PI vs. TH. (C) Incorporation plates, where filtrate from TH has been incorporated into media. Top, Rhizoctonia solani (RS) control; Bottom, RS + filtrate from TH (Courtesy of R. Scarselletti).

of Millipore sterilized test culture filtrate in small wells made in agar plates. The target organism can then be grown on this plate and its reaction to the presence of the culture filtrate recorded (Figure 2B). Alternatively, the sterile culture filtrate can be incorporated into the media and the test fungus can be grown on it. Changes in radial growth rate relative to the control can then be measured (Figure 2C). In all cases, identification of the substances that are active can be made by thin layer chromatography and gas-liquid chromatography with known standards.

IV. ANTAGONISM IN THE FIELD

A. Soil

The soil environment can be divided into two major ecological niches, the bulk soil that is uncolonized by plant roots, and the rhizosphere soil that exists around them. These environments exist more temporally than spatially, as during the normal growing cycle of crop plants almost all soil is infiltrated by plant roots, but after harvest and before the next planting almost all soil is bulk soil. Microbial antagonism of either dormant propagules or saprophytic mycelium in bulk soil is thus an important form of soil sanitation.

The prior colonization of a substrate by nonpathogenic microorganisms present in bulk soil is a major factor in preventing pathogens from becoming established on that same substrate.[13,15] Once established, these nonpathogenic colonizers must rely on vigorous competition, hyperparisitism, and antibiosis of potential plant pathogens to retain their grip on the substrate. The control of *Fusarium culmorum* has been achieved by exploiting this phenomena.[15]

Fusarium culmorum normally colonizes fresh wheat straw successfully when it is incorporated into soil, as often occurs when stubble is ploughed in. However, when the straw is allowed to stand in the field to become colonized by other saprophytes, it can no longer be colonized by *F. culmorum* when it is incorporated into the soil.

The utilization of antagonists that can attack dormant pathogen propagules in the soil by hyperparasitism has achieved a considerable degree of success in controlling several different diseases.[1-3,28,31,38,46,60] The invasion of oospores of *Phytophthora megasperma* v *sojae* by the *Pythium* sp. and *Aphanomyces euteiches* in soils by a number of different microorganisms has been reported by Sneh et al.[61] and Lumsden,[38] and the sclerotia of *Sclerotium sclerotiorum* are attacked by *Coniothyrium minutans* in soil, and can naturally decrease the numbers of viable sclerotia in the soil.[31,63]

An alternative to direct physical attack of mycelium or propagules is their lysis from a distance, a phenomenon that can be caused by either external antibiotics or toxins that are produced by other microorganisms, or by the germination-lysis phenomenon. The germination-lysis phenomenon refers to the stimulation of germination of fungal spores by the presence of nutrients, which is followed by the lysis of the germling through the action of the soil microbiota. Addition of nutrients to the soil in the form of organic amendments, root exudates, or fertilizers stimulates the dormant or semidormant pathogen propagules and these become activated and germinate. In the absence of a suitable host plant root and in the face of fierce nutrient competition the germling will lyse before it has a chance to produce any new resting structures. The cause of the lysis may be endogenous (autolysis) or it may be exogenous, by the extracellular enzymes or toxins produced from other soil microorganisms. The germination-lysis phenomena has been reported as affecting a great variety of soil-borne plant pathogens, including *Phymatotrichum omnivorum*, *Thelaviopsis basicola*, *Macrophomina phaseolina*, and *Fusarium solani*.[46]

A combination of all three mechanisms of antagonism is thought to operate in the development of disease suppressive soils. Several root pathogens exhibit cycles of severity followed by decline. Take-all, caused by the fungus *Gaeumannomyces graminis* var *tritici*, is a good example of plant disease that increases steadily in wheat monoculture, but after a certain number of years the disease begins to decline naturally.[55] This natural decline continues until the disease reaches barely detectable levels. It seems likely that the phenomenon is due to the establishment of a microflora in the soil that is antagonistic to the take-all fungus. It is possible to transfer this suppressiveness to disease conducive soils by large scale soil transfer from the decline area to one where disease is present.[54]

Similarly, biological control of *Phytophthora palmivora*, the causal agent of root rot of green papaya, has been successfully achieved by transferring rather smaller amounts of soils.[35] In Hawaii, papaya plants are very susceptible to *P. palmivora*, but the roots are only susceptible at the seedling stage. The control is achieved by drilling planting holes in the lava which are then filled with virgin soil that contains a natural microflora that is antagonistic to *Phytophthora*. The papaya seeds are then planted in the middle of the hole, and in this way the plants can become established before the *Phytophthora* can gain a stronghold. The control is thought to be due to a number of microorganisms rather than one specific one.

The role of organic amendments in controlling plant disease has been recognized for decades, and the mechanisms by which they achieve their effect can be many fold. Specific elements of the microbial population may be stimulated, and these may be antagonistic to certain disease causing organisms, or the stimulation can be much more general and involve almost all groups of organisms that are present in the soil. Some organic amendments can produce chemicals that are toxic to pathogens during their decomposition, or they can cause a change in the pH of the soil which then ceases to favor the pathogen. Similarly, there may be an alteration in the water holding capacity of the soil when organic materials are added, and this too may have a detrimental effect on the pathogen. This type of biological control of plant pathogens has been practiced by subsistence farmers for centuries, and before the advent of agrochemicals it provided the only form of disease control. There are thus many examples of disease control by soil organic amendments. Control of root rot caused by *Fusarium solani* f. sp. *phaseoli* has been achieved by the addition of chitin to the soil.[42] This amendment greatly increased the number of actinomycetes in the soil, and promoted the heterolysis of chitin walled fungi including *Fusarium*. Suppression of damping-off of peas, cucumber, and spinach caused by *Sclerotinia minor* has been achieved by the addition of composted sewage sludge to the soils, which because of the increased levels of organic matter and higher calcium levels, increased the activity of the saprophytic population of microorganisms.[39] Successful control of *Phytophthora cinnamomi* by the manipulation of the organic matter content of forest soils has been reported by Malajczuk[40] and Nesbitt et al.[44] Control of the late blight disease of potato caused by *Phytophthora infestans* has been achieved by amending soil with wheat straw that had been inoculated with *Trichoderma harzianum*.[56,57] And Mehrotra and Tiwari[41] successfully controlled the root rot of *Piper betle* caused by *Phytophthora parasitica* v. *parasitica* with corn straw as soil amendments.

B. The Rhizosphere

Garrett[25] described the rhizosphere as the outermost defense of the plant against pathogen attack of the roots. The composition of the rhizosphere has an important effect in determining the susceptibility or resistance of a plant to infection. The selective stimulation of a disease antagonistic population is known to be at least partly determined by the host[43] and is clearly exhibited in the formation of symbiotic root-fungus and root-bacteria associations.[9,27,64] A large number of saprophytic microorganisms compete for nutrients and space in the rhizosphere, and the mechanisms that determine the dominance of one type over another have already been discussed, i.e., antibiosis, competition, and hyperparsitism. A great number of saprophytic microorganisms compete for nutrients and space in the rhizosphere, and can give a significant degree of biological control of soil-borne plant pathogens.[56] The take-all fungus has been successfully ousted from the rhizosphere of wheat by its saprophytic relative *Phialophora radicicola*,[56] which occupies saprophytically the ecological niche in which its pathogenic cousin would have caused damage. Hyperparasitism also operates in the rhizosphere, and this can be clearly seen in electron microscopic studies of the region. Both bacteria and fungi can exhibit hyperparsitism. For example, the take-all fungus has been shown to be parasitized by bacteria in the rhizosphere of wheat.[56]

A number of actinomycetes and fungi that inhabit the rhizosphere have been shown to produce antibiotics, particularly members of the genera *Rhizobium*, *Bacillus*, and *Streptomyces*.[66] Some of the antibiotics have been identified. For example, a rhizosphere and phylloplane inhabiting isolate of *Pseudomonas flourescens* has been shown to produce two chlorinated phenyl pyrrole antibiotics.[29]

The importance of these types of compounds in antagonism in the rhizosphere has until recently been in doubt, but it has now been demonstrated that in the presence of sufficient organic substrate they are indeed produced and can play an important part in antagonism in the rhizosphere.[53] The exudates that plant roots release are rich in sugars, amino acids, and many other compounds, and provide a very suitable substrate for microbial colonization and the production of antibiotics.

C. Microbial Antagonism Within Plant Tissue

Microbial antagonism has often been considered as occurring mainly externally to the plant, in the phylloplane, rhizosphere, or in the bulk soil. However, some spectacular examples of antagonism between antagonist and pathogen within plant tissue exist. Control of *Armillaria mellea* root rot of citrus, in California, has been achieved by using a weak soil fumigant to destroy the normal biological balance of the soil and the root and stump debris of infected grubbed citrus trees. This treatment in itself does not destroy the pathogen, but significantly weakens it. In the absence of the normal microbiota, species of *Trichoderma* colonize the cellulosic remains of the trees and successfully antagonize the *Armillaria*, leading eventually to the eradication of the pathogen.[52] Control of the stump fungus *Fomes annosus*, a pathogen that causes considerable losses of pine and spruce trees globally, has been achieved by the precolonization of freshly cut tree stumps by the competitor *Peniophora gigantea*. The peniophora actively colonizes the woody tissues and by competition it prevents the pathogen from colonization. Furthermore, by the phenomenon of hyphal interference it eradicates the *Fomes* from any niches it has colonized.[50-52]

Another example of successful antagonism within host tissue occurs in the control of the silver leaf fungus (*Stereum purpureum*) with the antagonist *Trichoderma viride*. Infected trees can be treated by injection with *Trichoderma* impregnated dowels in to the heartwood. The dead, *Stereum* colonized wood become recolonized with the *Trichoderma*, and the trees can show complete recovery.[16,17]

D. Antagonism in the Phyllosphere

Competition for nutrients and space, the production of antibiotics, and hyperparasitism all play important roles in the antagonism of pathogens arriving and persisting in the phyllosphere.[6,7,21-23,36,37,58] Epiphytic microorganisms take up nutrients rapidly, resulting in rapid reductions in the amounts of nutrients available to pathogens.[11,22] Under these circumstances, spores of *Botrytis cinerea*, *Cladosporium herbarum*, and *Phoma betae* do not germinate or they germinate poorly, failing to give rise to infections. In these situations amino acids usually become limiting for growth before carbohydrates, and a relationship has been shown between the epiphytic microorganisms assimilation of $^{14}CO_2$ labeled amino acids and the degree of inhibition of pathogen spore germination which they caused.[7] Certain saprophytic bacteria are known to compete with pathogenic bacteria for nutrients. For example, the inhibition of *Xanthomonas oryzae* on rice shoots and *Erwinia amylovora* on fruit trees is due to the natural presence of *Erwinia herbicola* which reduces the amounts of nutrients available to the pathogens.[30,49] Fungal species are also reported as important in antagonism on the phylloplane, particularly species of *Trichoderma*.[56,62]

V. CONCLUSIONS

It appears that all of the mechanisms of antagonism that have been discussed, that is competition, hyperparasitism, and antibiosis, play an important role in the biological control of plant pathogens in the field. It is often difficult to extrapolate from laboratory experiments to the field to know exactly which mechanisms may be important in any individual interaction in any particular niche, but it is likely that at least two mechanisms are operating at any one time. However, great caution must be exercised when trying to exploit these natural interactions that have evolved over great lengths of time. It must be remembered that these features conferred greater environmental fitness upon their possessors in that particular niche only, and that attempting to exploit these phenomena in different situations will probably not result in their success, unless concomitant changes are made in the organisms that are being used. Thus, the use of conventional genetics and genetic engineering techniques to modify, exaggerate, or reduce certain features of the genome of the potential biological control agent will lead to the construction of new genotypes which, we hope, will be admirably suited to the new role in which they will find themselves. Understanding the interactions that occur between microorganisms is now insufficient, we must be able to modify these interactions too. If we do so then a great future awaits the exploitation of microbial antagonism to control plant disease.

REFERENCES

1. **Ayers, W. A. and Adams, P. B.**, Mycoparasitism of sclerotia of sclerotinia and sclerotium species by *Sporidesmium sclerotivorum, Can. J. Microbiol.*, 25, 17, 1979.
2. **Ayers, W. A. and Adams, P. B.**, Mycoparasitism and its application to biological control of plant diseases, in *Biological Control in Crop Production*, Papavizas, G. C., Ed., Allenheld, Totowa, N.J., 1981, 91.
3. **Ayers, W. A. and Lumsden, R. D.**, Mycoparasitism of oospores of *Pythium* and *Aphanomyces* spp. by *Hyphochytrium catenoides, Can. J. Microbiol.*, 23, 38, 1977.
4. **Baker, R.**, Mechanism of biological control of soil borne pathogens, *Annu. Rev. Phytopathol.*, 6, 263, 1968.
5. **Barnett, H. L.**, The nature of mycoparasitism by fungi, *Annu. Rev. Microbiol.*, 17, 1, 1963.
6. **Blakeman, J. P. and Fokkema, N. J.**, Potential for biological control of plant diseases on the Phylloplane, *Annu. Rev. Phytopathol.*, 20, 167, 1982.
7. **Blakeman, J. P. and Brodie, I. D. S.**, Competition for nutrients between epiphytic micro-organisms and germination of spores of plant pathogens on beetroot leaves, *Physiol. Plant Pathol.*, 10, 29, 1977.
8. **Boosalis, M. G.**, Hyperparasitism, *Annu. Rev. Phytopathol.*, 2, 363, 1964.
9. **Bowen, G. D. and Theodorou, C.**, Growth of ectomycorrhizal fungi around seeds and roots, in *Ectomycorrhizae their Ecology and Physiology*, Marks, G. C. and Kozlowski, T. T., Eds., Academic Press, New York, 1973, 107.
10. **Brian, P. W.**, The ecological significance of antibiotic production, in *Microbial Ecology*, Williams, R. E. O. and Spicer, L. L., Eds., Cambridge University Press, London, 1957, 1968.
11. **Brodie, I. D. S. and Blakeman, J. P.**, Competition for exogenous substrates *in vitro* by leaf surface micro-organisms and germination of conidia of *Botrytis cinerea, Physiol. Plant Pathol.*, 9, 227, 1976.
12. **Bruehl, G. W. and Lai, P.**, The probable significance of saprophytic colonisation of wheat straw in the field by *Cephalosporium gramineum, Phytopathology*, 58, 464, 1968.
13. **Bruehl, G. W., Millar, R. L., and Cunfer, B.**, Significance of antiobiotic production by *Cephalosporium gramineanum* to its saprophytic survival, *Can. J. Plant Sci.*, 49, 235, 1969.
14. **Clark, F. E.**, The concept of competition in microbial ecology, in *Ecology of Soil Borne Plant Pathogens*, Baker, K. F. and Snyder, W. C., Eds., California University Press, Berkeley, California, 1965, 339.
15. **Cook, R. J.**, Factors affecting saprophytic colonisation of wheat straw by *Fusarium roseum* f. sp. *cerealis* 'culmorum', *Phytopathology*, 60, 1672, 1970.
16. **Corke, A. T. K.**, Interactions between micro-organisms, *Ann. Appl. Biol.*, 89, 89, 1978.
17. **Corke, A. T. K. and Rishbeth, J.**, *Microbial Control of Pests and Plant Diseases 1970—1980*, Burges, H. D., Ed., Academic Press, New York, 1981, 914.
18. **Dennis, C. and Webster, J.**, Antagonistic properties of species group of *Trichoderma*. I. Production of non-volatile antibiotics, *Trans. Br. Mycol. Soc.*, 57, 25, 1971.
19. **Dennis, C. and Webster, J.**, Antagonistic properties of species group of *Trichoderma*. II. Production of volatile antibiotics, *Trans. Br. Mycol. Soc.*, 57, 41, 1971.
20. **Dennis, C. and Webster, J.**, Antagonistic properties of species group of *Trichoderma*. III. Hyphal interactions, *Trans. Br. Mycol. Soc.*, 57, 363, 1971.
21. **Fokkema, N. J.**, The influence of pollen on the development of *Cladosporium herbarum* in the phyllosphere of Rye, *Neth. J. Plant Pathol.*, 74, 159, 1968.
22. **Fokkema, N. J.**, Fungal leaf saprophytes beneficial or detrimental? in *Microbial Ecology of the Phylloplane*, Blakeman, J. P., Ed., Academic Press, London, 1981, 433.
23. **Fokkema, N. J. and Schippers, B.**, Phyllosphere versus shizosphere, as an environment for saprophytic colonisation, in 3rd Int. Symp. Microbiology of the Phylloplane, IAC, Wageningen, 1985.
24. **Fries, N.**, Effects of volatile organic compounds on the growth and development of fungi, *Trans. Br. Mycol. Soc.*, 60, 1, 1973.
25. **Garrett, S. D.**, *Pathogenic Root Infecting Fungi*, Cambridge University Press, London, 1970, 299.
26. **Gottlieb, D. and Shaw, P. D.**, Mechanism of action of antifungal antibiotics, *Annu. Rev. Phytopathol.*, 8, 371, 1970.
27. **Harley, J. L.**, *The Biology of Mycorrhiza*, Leonard Hill, London, 1959, 234.
28. **Hoch, H. C. and Abawi, G. S.**, Mycoparasitism of oospores of *Pythium ultimum* by *Fusarium merismoides, Mycologia*, 71, 621, 1979.

29. **Howell, C. R. and Stipancovic, R. D.,** Suppression of *Pythium ultimum* induced damping off of cotton seedlings by *Pseudomonas fluorescens* and its antibiotic Pyroluteorin, *Phytopathology,* 70, 712, 1980.

30. **Hsieh, S. P. Y. and Buddenhagen, I. W.,** Suppressing effects of *Erwinia herbicola* on infection by *Xanthomonas oryzae* and on symptom development of rice, *Phytopathology,* 64, 1182, 1974.

31. **Huang, H. C.,** Importance of *Coniothyrium minitans* in survival of sclerotia of *Sclerotinia sclerotium* in wilted sunflower, *Can. J. Bot.,* 55, 289, 1977.

32. **Hutchinson, S. A.,** Biological activity of volatile fungal metabolites, *Annu. Rev. Phytopathol.,* 11, 223, 1973.

33. **Ikediugwu, F. E. O. and Webster, J.,** Antagonism between *Coprinus heptemerus* and other coprinus fungi, *Trans. Br. Mycol. Soc.,* 54, 181, 1970.

34. **Ikediugwu, F. E. O. and Webster, J.,** Hyphal interference by *Peniophora gigantea* against *Heterobasidium annosum, Trans. Br. Mycol. Soc.,* 54, 307, 1970.

35. **Ko, W. H.,** Biological control of seedlings root rot of Papaya caused by *Phytophthora palmivora, Phytopathology,* 61, 780, 1971.

36. **Kranz, J.,** Hyperparasitism of biotrophic fungi, in *Microbial Ecology of the Phylloplane,* Blakeman, J. P., Ed., Academic Press, London, 1980, 502.

37. **Kranz, J.,** Ecology of mycoparasitism, in *The Fungal Community and its Role in the Ecosystem,* Wicklow, B. J. and Carroll, G. W., Eds., Marcel Dekker, New York, 1980.

38. **Lumsden, R. D.,** Hyperparasitism for the control of plant pathogens, in *Handbook of Pest Management in Agriculture,* Vol. 1, Pimentel, D. and Hanson, A. A., Eds., CRC Press, Boca Raton, 1980.

39. **Lumsden, R. D.,** The role of micro-organisms in sustainable agriculture, *2nd Int. Conf. Biological Agriculture and Horticulture,* Lopez-Real, J. M. and Hodges, R. D., Eds., Academic Press, London, 1986.

40. **Malajczuk, N.,** Biological suppression of *Phytophthora cinnamoni* in Eucalyptus and Avocadoes in Australia, in *Soil Borne Plant Pathogens,* Schippers, B. and Gams, W., Eds., Academic Press, London, 1979, 635.

41. **Mehrotra, R. S. and Tiwari, D. P.,** Organic amendments and control of foot root of *Piper betle* caused by *Phytophthora parasitica* var. *Piperina, Ann. Microbiol. Inst. Pasteur,* 127(A), 415, 1976.

42. **Mitchell, R. and Alexander, M.,** Microbiological processes associated with the use of Chitin for biological control, *Proc. Soil Sci. Soc. Am.,* 26, 556, 1962.

43. **Neal, J. L., Jr., Atkinson, T. G., and Larson, R. I.,** Changes in the rhizosphere microflora of spring wheat induced by disomic substitution of a chromosome, *Can. J. Microbiol.,* 16, 153, 1970.

44. **Nesbitt, H. L., Malajczuk, N., and Glenn, A. R.,** Effect of organic matter on the survival of *Phytophthora cinnamoni* Rands in soil, *Soil Biol. Biochem.,* 11, 133, 1979.

45. **Odum, E. P.,** *Fundamentals of Ecology,* W. B. Saunders, Philadelphia, 1953, 384.

46. **Papavizas, G. C. and Lumsden, R. D.,** Biological control of soil borne fungal propagules, *Annu. Rev. Phytopathol.,* 189, 389, 1980.

47. **Park, D.,** *Antagonism — The Background of Soil Fungi,* Liverpool University Press, Liverpool, England, 1960, 148.

48. **Park, D.,** The importance of antibiotics and inhibiting substances in soil biology, in *Soil Biology,* Burges, A. and Raw, F., Eds., Academic Press, London, 1967, 435.

49. **Riggle, J. H. and Klos, E. J.,** Relationship of *Erwinia herbicola* to *E. Amylovora, Can. J. Bot.,* 50, 1677, 1972.

50. **Rishbeth, J.,** Stump protection against *Fomes annosus.* III. Inoculation with *Peniophora gigantea,* *Ann. Appl. Biol.,* 52, 63, 1963.

51. **Rishbeth, J.,** Stump inoculation: a biological control of *Fomes annosus,* in *Biology and Control of Soil Borne Plant Pathogens,* Bruehl, G. W., Ed., Arondel Phytopathology Society, St. Paul, Minn., 1975, 216.

52. **Rishbeth, J.,** Modern aspects of biological control of *Fomes* and *Armillaria, Eur. J. For. Pathol.,* 9, 331, 1979.

53. **Rothock, C. S. and Gottlieb, G.,** Role of antibiosis in antagonism of *Streptomyces hygroscopicus v. geldans* to *Rhizoctonia solani* in soil, *Can. J. Microbiol.,* 30, 1440, 1984.

54. **Shipton, P. J., Cook, R. J., and Sihon, J. W.,** Occurrence and transfer of a biological factor in soil that suppress take all of wheat in Eastern Washington, *Phytopathology,* 63, 511, 1973.

55. **Shipton, P. J.,** Monoculture and soil borne plant pathogens, *Annu. Rev. Phytopathol.,* 15, 387, 1977.

56. **Singh, J. S.,** Biological control of *Phytophthora infestans,* Ph.D. thesis, University of London, London, England, 1986.

57. **Singh, J. S. and Faull, J. L.,** Biological control of *Phytophthora infestans,* The role of micro-organisms in a sustainable agriculture, in *2nd Int. Conf. Biological Agriculture and Horticulture,* Lopez-Real, J. M. and Hodges, R. D., Eds., Academic Press, London, 1986.

58. **Singh, J. S. and Faull, J. L.,** Hyperparasitism and biological control of late blight fungus, in *4th Int. Symp. Microbiology Phylloplane,* Wageningen, Holland, September, 1985, in press.

59. **Skidmore, A. M.,** Interactions in relation to biological control of plant pathogens, in *Microbiology of Aerial Plant Surfaces,* Dickinson, C. H. and Preece, T. F., Eds., Academic Press, London, 1976, 507.

60. **Sneh, B., Holdaway, B. F., Hooper, G. R., and Lockwood, J. L.,** Germination lysis as a mechanism for biological control of *Thielaviopsis basicola* pathogenic on soybean, *Can. J. Bot.,* 54, 1499, 1976.

61. **Sneh, B., Humble, S. J., and Lockwood, J. L.,** Parasitism of oospores of *Phytophthora megasperma* var. *sojae, P. cactorum, Pythium* sp. and *Aphanomyces euteiches* in soil by comycetes, Chytridiomycetes, Hyphomycetes, actinomycetes and bacteria, *Phytopathology,* 67, 622, 1977.

62. **Sundheim, L.,** Use of hyperparasites in biological control of biotrophic pathogens, in *4th Int. Symp. Microbiology of the Phyllosphere,* Fokkema, N. J. and Van den Heuvel, J., Eds., Cambridge University Press, London, 1985.

63. **Turner, G. J. and Tribe, H. T.,** On *Coniothyrium minitans* and its parasitism of *Sclerotinia* sp., *Trans. Br. Mycol. Soc.,* 66, 97, 1976.

64. **Vincent, J. M.,** Environmental factors in the fixation of nitrogen by the legume, in *Soil Nitrogen,* Bartholomew, W. B., and Clark, F. E., Eds., American Society of Agronomy, Madison, Wis., 1965, 384.

65. **Waksman, S. A.,** *Soil Microbiology,* John Wiley & Sons, New York, 1952, 356.

66. **Weinhold, A. R. and Bowman, T.,** Selective inhibition of the potato scab pathogen by antagonistic bacteria and substrate influence on antibiotic production, *Plant Soil,* 28, 12, 1968.

Chapter 13

BIOCONTROL OF STORAGE MOLD DISEASES OF SEED

John Tuite

TABLE OF CONTENTS

I. INTRODUCTION

Storage diseases caused by microorganisms are characterized by a sufficient loss in quality to reduce end use or grade of seeds. Generally, storage diseases are controlled by manipulation of the environment. For long-term storage, grain moisture is reduced and maintained at a safe level, e.g., interseed relative humidity to below 70%. Lowering or at least stabilization of grain temperature is desirable because uniform temperatures in grain prevents localized moisture build-up through moisture migration. Grain drying, conditioning, and storage has been the subject of many extensive treatments[6,8,19,56,60] and will only be treated indirectly. This review will consider primarily the use of germplasm resistant to mold attack and damage as a means to increase grain storability. Resistance to storage molds is proposed not as a sole method of control but as an adjunct to conventional environmental methods.[60] Also considered is the employment of antagonistic microorganisms to decrease the activities of harmful storage molds.

The search for and the utilization of genotypes resistant (some prefer to designate the condition as less susceptible) to storage molds is just beginning, as indicated by the relatively few published reports. These reports, and what may be the future directions of research, will be discussed.

Lessening of mold-induced losses by the utilization of better germplasm may be sought primarily in two ways. The first and most apparent is to seek seed resistance to one or more of the following: mold infection, colonization, and sporulation. In addition, some have emphasized screening for resistance of the seed to mycotoxin production. Regardless of whether resistance to fungal deterioration or mycotoxin production is sought, another dimension, particularly with corn and other field crops, must be considered. Corn kernels in the U.S. and other areas of high mechanization, and certain other grains are subject to considerable physical abuse when harvested and stored. Therefore, in addition to, or some might argue in lieu of mold resistance, resistance of the kernel to harvest, handling, and drying damage should be pursued. Practical experience and laboratory tests[20,23,44,50,54] indicate that mold activity is greatly increased by physical damage to seed. Therefore, genotypes resistant to physical damage[18,22,28,42,45,55,68] offer a substantial opportunity to lessen mold-induced losses and also make the grain less likely to generate conditions where dust explosions will occur.

Another consideration is resistance to insect attack. Because insects and mites favor mold development, resistance to insects would benefit mold resistance.[32]

II. MAINTENANCE OF SEED VIABILITY

Reports[2,17,24,35,37,40] on a variety of seed crops indicated that certain lines, inbreds, or cultivars maintained their viability longer than other germplasms, and longevity is a breedable trait.[2,48] These storage studies generally were at seed moistures in equilibrium with relative humidities of <70% where fungi were unable to grow, or were at high temperatures where seed deterioration can be rapid without microbial action. In these circumstances maintenance of high germination can not be ascribed to storage mold resistance. They suggest, however, that these genotypes would be good storers because molds grow less vigorously on living seeds than on dead seeds.

III. RESISTANCE TO STORAGE MOLDS

An early indication of resistance to storage molds was demonstrated by Mayne et al.,[30] who found that a cotton seed lot with an impermeable seed coat was invaded less by *Aspergillus flavus,* and had much less aflatoxin, than a line with a permeable seed coat.

Work by Moreno-Martinez and Christensen[36] was perhaps the first extensive demonstration that the maintenance of seed viability in storage was a manifestation of resistance to storage mold growth. Later, Moreno et al.[37] evaluated storability as determined by seed viability of inbreds, single, and double crosses of corn recommended in Mexico. In the earlier work[36] they noted that corn entries having high viability had less visible mold and generally had fewer infected kernels at 85% RH. Highly viable seed had less *Aspergillus candidus* infection, a storage mold associated with decreased seed viability. Their results, that different hand harvested corn genotypes react differently to storage molds, was confirmed and extended by Cantone et al.[7] Mold development of inoculated corn kernels, stored at relative humidities of 85, 88, and 91%, was measured by several techniques: plating of surface disinfected kernels (percentage kernel infection), dilution technique (numbers of fungal propagules), visible estimates of mold sporulation, and ergosterol content (a fungal metabolite). Correlations were generally high between these measures, but since estimates of visible mold could be done rapidly and without much equipment and training it could be a routine screening technique. Percentage of infected kernels was not as discriminating as the other criteria unless sampling was done early in storage. Most kernels became invaded with time of storage, and resistance to storage molds with these genotype was relative; this was also indicated by the other measures of fungal development. Correlations of seed germination with fungal evaluation were poor in most cases, and may indicate that under the conditions and length of storage, loss in seed germination was slight except when the kernels were inoculated with *A. flavus* and stored at conditions for rapid mold growth. The authors concluded that resistance appeared to be against all three of the important storage molds they tested: the *Aspergillus glaucus* group, *A. flavus* group, and three species of *Penicillium*. However, Ilag and Juliana,[16] in their studies with heat-killed rice seed, found differences on resistance to sporulation between *A. flavus* and *A. parasiticus*. Cantone et al.[7] found the resistance of corn inbreds to storage molds generally inferior to hybrids, and the range of differences among four midwest corn hybrids to be modest but significant. Tuite et al.[61] suggested that resistance in their corn genotypes existed in the pedicel, pericarp, and the germ. Other tests with corn for resistance to *A. flavus*, that may relate to resistance to mold growth in storage are, in vitro tests in which the seed is placed on agar media containing 6% $NaCl^2$ or moistened filter paper,[21] and kernels are rated for sporulation after a short (usually 7 days) incubation. The results of these tests have not been compared to storage tests. Yao[67] used sucrose in a water agar medium at concentrations sufficient to inhibit germination but not enough to kill the corn during a 2 week test period at 13°C. Resistance was evaluated visually and by dilution counts. Preliminary results indicated a correlation with the results of laboratory storage tests.

Onesirosan,[41] in storage tests of cowpea inoculated and uninoculated with members of the *A. glaucus* group, found differences in resistance to *A. glaucus* and the *Penicillium* spp. (the latter developed spontaneously). Genotypes that had a smooth seed coat had less visible mold, odor, seed infection, and loss in seed germination than those with wrinkled coats. Susceptibility also appeared to be associated with a high incidence of cracks in the seed coat or insect punctures.

Mixon and Rogers[33] identified a number of peanut lines that had less seed colonization by *A. flavus* as observed visually after inoculation and storage for 7 days at 25°C in a saturated humidity with a prior moisture adjustment. These lines are screened to obtain resistance to *A. flavus* in "the field and in the drying process immediately after harvest."[33] Performance of these two lines in regard to aflatoxin production in storage was not satisfactory according to Wilson et al.[66] They have been registered [34] and are available for genetic studies and incorporation into varieties with more acceptable crop yields.

IV. RESISTANCE TO MYCOTOXIN PRODUCTION

The first reported success of efforts to prevent aflatoxin production by host resistance in peanuts was in 1967.[57] Subsequently, there have been conflicting reports on resistance.[10,39,43,66] One of the difficulties in assessing these reports is the different techniques that were employed, often given with little detail and with little statistical treatment. For example, prestorage treatment of the kernels included unknown treatments, no treatments, and autoclaved; isolates of *A. flavus* were different in all studies; the amount of inoculum included an unknown amount, 30,000 spores/g, or 80,000 spores/g; moisture content was usually unreported, "rehydrated", or stored at a definite relative humidity but without mention of initial moisture; storage temperatures were unreported, or 23°C to 30°C; the condition of the kernels was not described, healthy, or sound; and the method used for harvesting was unknown or by machine. Frequently, how the samples were grown and stored prior to the test is unknown. Because of all these variables the significance of resistance to aflatoxin production in peanuts is obscure. It appears that resistance does occur, but it is variable, is strongly compromised by damage to the testa, and resistance to aflatoxin production is not accompanied by resistance to mold growth or sporulation.[31,43] Kushalappa et al.[26] found pod resistance to *A. flavus* variable. They suggested that resistance may be caused by antagonistic organisms, and as a consequence, be too variable to be used in breeding programs.

Partial resistance to aflatoxin production, but not infections by *A. flavus,* was found in a few cultivars of cowpea heavily inoculated and stored under optimum conditions for *A. flavus*. A line of cotton seed with an impermeable seed coat was shown by Mayne et al.[30] to have less infection with *A. flavus* and other fungi, and also supported substantially less aflatoxin production in storage. Shotwell[55] found differences in 16 soybean varieties.

Resistance to aflatoxin production has been extensively sought in corn, primarily prior to harvest. The results have been disappointing because of the variable amounts of aflatoxins at different geographical locations and from 1 year to the next.[9,27,65] Partly because of erratic field results, in vitro kernel tests or kernel homogenates of various maturities in culture media were devised.[1,38,62,64] The results are inconclusive, but the tests have not been extensively evaluated, in part because of the problem of relating the in vitro tests to variable field-based data. It is also not known how well these in vitro tests relate to aflatoxin production in storage. In vitro resistance to zearalenone[52] has been demonstrated.

Lines of rice, heated at 60°C, moistened, and inoculated with *A. parasiticus* and *A. flavus* were evaluated by Ilag and Juliana[16] for resistance to sporulation and aflatoxin production in storage. The amount of sporulation did not correlate closely with aflatoxin production. In addition, neither sporulation nor aflatoxin production by the two species corresponded. One line, however, supported little aflatoxin with both species. Surprisingly, lines with thicker pericarps tended to support greater aflatoxin production; a suggested reason was the greater food source in the thicker pericarp.

Substantial differences in the amount of aflatoxin B_1 and G_1 produced in 16 cultivars of soybean were reported by Shotwell et al.[53] when nonsterilized cracked seed were separately inoculated by isolates of *A. flavus* and *A. parasiticus*. Screening for resistance to zearalenone production by 3 isolates of *Fusarium graminearum* was performed on 14 corn inbreds and 4 single cross hybrids at 45% moisture content.[52] Visible mold growth did not correlate with zearalenone amounts. Inbreds generally contained smaller amounts of zearalenone than hybrids, but there was considerable variation within the two groups. Several inbreds and two single crosses appeared to inhibit zearalenone production.

The utility of resistance to mycotoxin production by itself is uncertain for two reasons. If a genotype suppresses production of a mycotoxin but not fungal growth, marketing problems remain. In addition, other metabolic products produced by the mycotoxigenic fungus may be toxic or have negative nutritive or palatability effects.

V. DAMAGE AND ITS EFFECT ON MOLD GROWTH AND RESISTANCE TO STORAGE MOLDS.

Physical damage to the seed enhances mold growth as measured by microbiological techniques and by CO_2 production. In addition, it is well documented that broken seed and dust decreased air flow with a concomitant increase in management problems. Qasem and Christensen,[44] Seitz et al.,[50] and Steele et al.[54] all found that damage to the corn kernel, particularly to the germ or its vicinity, greatly increased fungal activity in storage. In fact, kernels without apparent damage that were combined-harvested produced more CO_2 than hand-shelled corn.[20]

Damage to the peanut seeds resulted in a dramatic increase in aflatoxin in stored peanuts.[31,46] Because of the considerable enhancement of fungal growth by physical damage, does such damage in corn nullify kernel resistance? Despite significant increases in sporulation of *Penicillium* caused by a variety of injuries, statistical differences in sporulation remained, in most cases, between resistant and susceptible genotypes.[61] Since not every kernel would be damaged during combine-harvesting, as they were in these laboratory tests, it appears that selection and/or development of resistant hybrids may be valuable. However, because of the significant effects of kernel damage on mold growth, resistance to physical damage can not be neglected. Mehan and McDonald[31] stated "the maximum advantage than can be derived from groundnut cultures resistant to *A. flavus* will only occur if the seed testa is not damaged." Generally, improvement of corn hybrids has been directed toward traits such as yield, maturity, standability, adaptability, test weight, and resistance to insects and diseases in the field. Characteristics such as storage mold resistance, storage insect resistance, and milling and handling qualities have been largely ignored. In fact, as Halloin (11) summarizes, breeding efforts have often been away from those traits that would confer fungal resistance. There appears to be differences in resistance to physical damage in existing corn hybrids. Koehler[23] found corn hybrids to differ in their ability to resist pericarp damage inflicted by shelling. There was a significant difference between two commercial hybrids in the amount of kernel damage when combine harvested.[45] Mahmoud and Kline[28] and Kline[22] reported wide differences in pericarp thickness in combine harvested corn in Iowa, and that the thickness was inversely proportional to the number of hairline cracks and minor pericarp damage. Pericarp thickness appeared to have little effect on preventing crushing and breaking of kernels.[28] Varietal differences in the cracking of the seed coat in soybeans were found by Yasue and Kinomura,[68] but there was an inverse correlation of seed coat thickness and cracking.

Paulsen et al.,[42] in a 4 year study of U.S. Corn Belt genotypes (inbreds and hybrids), found that a high drying temperature (60°C) affected susceptibility to breakage more than the differences between genotypes, but there was a significant difference between genotypes. Johnson and Russell[18] demonstrated the potential for selection of corn genotypes that resist kernel breakage in a Stein mill test. They used a visual rating of the endosperm for rapid evaluation of kernel hardness. They concluded that a flint type kernel is less liable to break, but selections for resistance to breakage may result in smaller seed. They believed that these selections would not have to result in yield reductions. Stroshine et al.[55] demonstrated significant differences in susceptibility to breakage within Corn Belt genotypes, but found that some genotypes reacted differently after high temperature drying giving changed rankings.

VI. HISTOLOGICAL AND CHEMICAL STUDIES OF STORAGE MOLD INVASION

Most histopathological studies of seed invasion by fungi have dealt with preharvest invasion. Recently, Tsuruta et al.,[59] using scanning electron microscopy, and Schans et al.,[47] using fluorescent microscopy, obtained information about routes of penetration of storage molds in corn kernels and rape seed, respectively. In rape seed an apparent resistant reaction to invasion was noted. These kinds of studies, if enlarged to include stored resistant and susceptible material, should locate where resistance occurs. They may also indicate the nature of the resistance, whether it is morphological or chemical, or more likely, a combination of the two. As mentioned previously, Onesirosan[41] believes the nature of the testa in cowpea determines resistance to storage molds. Harman and others[12-14] have demonstrated, by fungal isolation from seed tissues and chemical procedures, that chemical action by storage molds may be significant in the loss of seed viability in some pathogen-host interaction and not in others. Other seed entirely resists invasion by storage fungi.[25] Both chemical and histological studies should assist in developing a more rational and perhaps a more efficient selection of resistance to storage molds.

VII. CONTROL OF HARMFUL STORAGE MICROORGANISMS BY THE USE OF ANTAGONISTIC MICROORGANISMS

Employment of antagonistic microorganisms to inhibit significant storage fungi is seriously complicated by marketing and quality requirements. Grain and its products are purchased, in part, on the appearance of wholesomeness, soundness, freedom from odors, and signs and symptoms of microbial growth. For these reasons, ensiled fermented grain is only used for feed on the farm. Yeasts and bacteria combine with high moisture grain to develop a semi-anaerobic system in sealed bins. They prevent the growth of harmful aerobic storage fungi, but produce a sour odor unacceptable for merchandizing. The ensiled product is, however, fully suitable for feeding to cattle. Several different fungi inhibit the production of aflatoxin in stored grain,[3-5,15,29,63] including nonaflatoxigenic strains of *A. flavus*,[58] but their own growth is likely to be undesirable and may not always be reliable or predictable.[51] Other fungi, although inhibitory, may require high moisture and aerobic conditions. Sauer[46] has reported in preliminary studies some success with certain species of the *A. glaucus* group in lessening the growth and concomitant damage of more aggressive and more harmful microorganisms without visibly affecting grain quality. Other ''unobjectionable'' organisms may be found that can inhibit highly toxic or mycotic organisms. Their use seems more applicable if there is a recurring and a serious problem with toxic fungi, if grain conditioning equipment is unavailable, or climate makes dry storage difficult to obtain and maintain. Such circumstances may likely occur in developing countries with feed grains.

A possible benefit from studies on antagonistic fungi is that they may reveal compounds inhibitory to storage mold growth or mycotoxin production.

VIII. SUMMARY AND RECOMMENDATIONS

Antagonistic organisms must inhibit significant storage organisms without themselves adversely affecting grain quality or end use. Because this latter constraint is difficult to overcome, it appears that utilization of genetic resistance is the more attractive path to decreasing storage diseases. Storage mold resistance, although not likely to be absolute, would supplement existing environmental control methods. Resistance to storage molds appears relative, therefore test procedures should be not so severe that real and useful differences are obscured. It also seems necessary to evaluate germplasm grown at different locations for several years, to judge resistance. For resistance to be achieved and commercially feasible the following requirements appear necessary:

1. Resistance is stable and effective against most, if not all, storage pathogens.
2. Resistance is not associated with negative factors such as poor yield, milling and handling properties, and antinutritive factors. Other considerations may be appropriate depending on the seed and its end use.

3. Resistance is not readily overcome by normal harvest and postharvest operations. In fact, resistance to these predisposing factors should be actively sought in crops subjected to significant physical abuse.
4. Resistance to mycotoxin production should be evaluated as an important attribute. Conceivably, a genotype could restrict sporulation but support unacceptable levels of mycotoxin production.
5. Resistance to storage molds may be simply tested by a standardized procedure. An in vitro test that is relevant to storage performance would be highly desirable.

Knowledge of the nature and location of resistance in the seed would aid in the screening for resistance to storage molds. Resistance to insect attack would also be highly desirable as grain insects significantly promote mold growth.

ACKNOWLEDGMENT

Thanks to Roy Curtis for a critical review of the manuscript.

REFERENCES

1. **Adams, N. J., Scott, G. E., and King, S. B.,** Quantification of *Aspergillus flavus* growth on inoculated excised kernels of corn genotypes, *Agron. J.,* 76, 98, 1984.
2. **Agrawal, P. K., Patil, R. B., Dadlani, M., and Singh, D.,** Effect of relative humidity and temperature on germination of seeds of two F_1 sorghum hybrids and their parents during storage, *J. Seed Technol.,* 6, 31, 1981.
3. **Ashworth, L. J., Schroeder, H. W., and Langley, B. C.,** Aflatoxins: environmental factors governing occurrence in Spanish peanuts, *Science,* 148, 1228, 1965.
4. **Boller, R. A. and Schroeder, H. W.,** Influence of *Aspergillus chevalieri* on production of aflatoxin in rice by *Aspergillus parasiticus, Phytopathology,* 63, 1507, 1973.
5. **Boller, R. A. and Schroeder, H. W.,** Influence of *Aspergillus candidus* on production of aflatoxin in rice by *Aspergillus parasiticus, Phytopathology,* 64, 121, 1974.
6. **Brooker, D. B., Bakker-Arkema, F. W., and Hall, C. W.,** *Drying Cereal Grains,* AVI Publishing, Westport, Conn., 1974.
7. **Cantone, F. A., Tuite, J., Bauman, L. F., and Stroshine, R.,** Genotypic differences in reaction of stored corn kernels to attack by selected *Aspergillus* and *Penicillium* spp., *Phytopathology,* 73, 1250, 1983.
8. **Christensen, C. M.,** *Storage of Cereal Grains and their Products,* American Association of Cereal Chemists, St. Paul, Minn., 1982.
9. **Davis, N. D., Currier, C. G., and Diener, U. L.,** Response to corn hybrids to aflatoxin formation by *Aspergillus flavus, Ala. Agric. Exp. Stn. Bull.,* 475, 1985.
10. **Doupnik, B., Jr.,** Aflatoxins produced on peanut varieties previously reported to inhibit production, *Phytopathology,* 59, 1554, 1969.
11. **Halloin, J. M.,** Deterioration resistance mechanisms in seeds, *Phytopathology,* 73, 335, 1983.
12. **Harman, G. E.,** Mechanisms of seed infection and pathogenesis, *Phytopathology,* 73, 326, 1983.
13. **Harman, G. E. and Pfleger, F. L.,** Pathogenicity and infection sites of *Aspergillus* species in stored seeds, *Phytopathology,* 64, 1339, 1974.
14. **Harman, G. E. and Nash, G.,** Deterioration of stored pea seed by *Aspergillus ruber:* evidence for involvement of a toxin, *Phytopathology,* 62, 209, 1972.
15. **Horn, B. W. and Wicklow, D. T.,** Factors influencing the inhibition of aflatoxin production in corn by *Aspergillus niger, Can. J. Microbiol.,* 29, 1087, 1983.
16. **Ilag, L. L. and Juliana, B. O.,** Colonisation and aflatoxin formation by *Aspergillus* spp. on brown rices differing in endosperm properties, *J. Sci. Food Agric.,* 33, 97, 1982.
17. **James, E., Bass, L. N., and Clark, D. C.,** Varietal differences in longevity of vegetable seeds and their responses to various storage conditions, *Am. Soc. Hort. Sci. Proc.,* 91, 521, 1967.
18. **Johnson, D. Q. and Russell, W. A.,** Genetic variability and relationships of physical grain-quality traits in the BSSS population of maize, *Crop Sci.,* 22, 805, 1982.

19. **Justice, O. L. and Bass, L. N.,** Principles and Practices of Seed Storage, *Agric. Handbook No. 506,* U.S. Department of Agriculture, Washington, D.C., 1978.

20. **Kalbasi-Ashtari, A., Bern, C. J., and Kline, G. L.,** Effect of internal and external damage on deterioration rate of shelled corn, Paper No. 79-3038, *American Society of Agricultural Engineers, St. Joseph, Mich.,* 1979.

21. **King, S. B. and Wallin, J. R.,** Methods for screening corn for resistance to kernel infection and aflatoxin production by *Aspergillus flavus, South. Coop. Ser. Bull.,* 279, 1983.

22. **Kline, G. L.,** Mechanical damage levels in shelled corn from farms, Paper No. 73-331, *American Society of Agricultural Engineers, St. Joseph, Mich.,* 1973.

23. **Koehler, B.,** Pericarp injuries in seed corn: prevalence in dent corn and relation to seedling blights, *Illinois Univ. Agric. Exp. Sta. Bull.,* 67, 1957.

24. **Kueneman, E. A.,** Genetic control of seed longevity in soybeans, *Crop Sci.,* 23, 5, 1983.

25. **Kulik, M. M.,** Susceptibility of stored vegetable seeds to rapid invasion by *Aspergillus amstelodami* and *A. flavus* and effect on germinability, *Seed Sci. Technol.,* 1, 799, 1973.

26. **Kushalappa, A. C., Bartz, J. A., and Norden, A. J.,** Susceptibility of pods of different peanut genotypes to *Aspergillus flavus* group fungi, *Phytopathology,* 69, 159, 1979.

27. **Lillehoj, E. B.,** Variability in corn hybrid resistance to preharvest aflatoxin contamination, *J. Am. Oil Chem. Soc.,* 58, 970A, 1981.

28. **Mahmoud, A. R. and Kline, G. L.,** Effect of pericarp thickness on corn kernel damage, Grain Damage Symp., Ohio State University, Columbus, Ohio, 1972.

29. **Malani, R., Venkitasubramanian, T. A., and Mukerji, K. G.,** Microbial interactions and aflatoxin production, *Proc. Symp. Mycotoxin in Food and Feed,* Bilgrami, K. S., Prasad, T., and Sinha, K. K., Eds., Allied Press, Bhagalpur, India, 1983, 297.

30. **Mayne, R. Y., Harper, G. A., Franz, A. O., Jr., Lee, L. S., and Goldblatt, L. A.,** Retardation of the elaboration of aflatoxin in cotton seed by impermeability of the seed coats, *Crop Sci.,* 9, 147, 1964.

31. **Mehan, V. K. and McDonald, D.,** Research on the aflatoxin problem in groundnut at ICRISAT, *Plant Soil,* 79, 255, 1984.

32. **Mills, J. T.,** Insect-fungus associations influencing seed deterioration, *Phytopathology,* 73, 330, 1983.

33. **Mixon, A. C. and Rogers, K. M.,** Peanut accessions resistant to seed infection by *Aspergillus flavus, Agron. J.,* 65, 560, 1973.

34. **Mixon, A. C. and Rogers, K. M.,** Registration of *Aspergillus flavus* — resistant peanut germplasms, *Crop Sci.,* 15, 106, 1975.

35. **Moreno-Martenez, E.,** Effecto de los hongos de almacen sobre la viabilidad de las semillas de maiz y soya, *Bol. Soc. Mex. Mic.,* 13, 195, 1979.

36. **Moreno-Martinez, E. and Christensen, C. M.,** Differences among lines and varieties of maize in susceptibility to damage by storage fungi, *Phytopathology,* 61, 1498, 1971.

37. **Moreno-Martinez, E., Morones-Reza, R., and Gutierrez-Lombardo, R.,** Diferencias entre lineas, cruzas simples y dobles de maiz en su susceptibilidad al dano por condiciones adversas de almacenamiento, *Turrialba,* 28, 233, 1978.

38. **Nagarajan, V. and Bhat, R. V.,** Factor responsible for varietal differences in aflatoxin production in maize, *J. Agric. Food Chem.,* 20, 911, 1972.

39. **Nagarajan, V. and Bhat, R. V.,** Aflatoxin production in peanut varieties by *Aspergillus flavus* Link. and *Aspergillus parasiticus* Speare., *Appl. Microbiol.,* 25, 319, 1973.

40. **Neal, N. P. and Davis, J. R.,** Seed viability of corn inbreds as influenced by age and conditions of storage, *Agron. J.,* 48, 383, 1956.

41. **Onesirosan, D. T.,** Effect of seed coat type on the susceptibility of cowpeas to invasion by storage fungi, *Seed Sci. and Technol.,* 10, 631, 1982.

42. **Paulsen, M. R., Hill, L. D., White, D. G., and Sprague, G. F.,** Breakage susceptibility of corn belt genotypes, *Trans. ASAE,* 26, 1830, 1983.

43. **Priyadarshini, E. and Tulpule, P. G.,** Relationship between fungal growth and aflatoxin production in varieties of maize and groundnut, *J. Agric. Food Chem.,* 26, 249, 1978.

44. **Qasem, S. A. and Christensen, C. M.,** Influence of various factors on the deterioration of stored corn by fungi, *Phytopathology,* 50, 703, 1960.

45. **Racop, E., Stroshine, R., Lien, R., and Notz, W.,** A comparison of three combines with respect to harvesting losses and grain damage, Paper No. 84-3016, *American Society of Agricultural Engineers, St. Joseph, Mich.,* 1984.

46. **Sauer, D. B.,** Inhibition of damaging storage fungi by *Aspergillus glaucus* in grain, *Phytopathology,* 75 (Abstr.), 1378, 1985.

47. **Schans, J., Mills, J. T., and Van Caeselle, L.,** Fluorescence microscopy of rape seeds invaded by fungi, *Phytopathology,* 12, 1582, 1982.

48. **Scott, G. E.,** Improvement for accelerated aging response of seed in maize populations, *Crop Sci.,* 21, 41, 1981.

49. **Seenappa, M., Keswani, C. L., and Kundya, T. M.,** *Aspergillus* infection and aflatoxin production in some cowpea (*Vigna unguiculata* (L.) Walp) lines in Tanzania, *Mycopathologia,* 83, 103, 1983.
50. **Seitz, L. M., Sauer, D. B., and Mohr, H. E.,** Storage of high-moisture corn: fungal growth and dry matter loss, *Cereal Chem.,* 59, 100, 1982.
51. **Seitz, L. M., Sauer, D. B., Mohr, H. E., and Aldis, D. F.,** Fungal growth and dry matter loss during bin storage of high-moisture corn, *Phytopathology,* 59, 9, 1982.
52. **Shannon, G. M., Shotwell, O. L., Lyons, A. J., White, D. G., and Garcia-Aquirre, G.,** Laboratory screening for zearalenone formation in corn hybrids and inbreds, *J. Assoc. Off. Anal. Chem.* 63, 1275, 1980.
53. **Shotwell, O. L., Vandegraft, E. E., and Hesseltine, C. W.,** Aflatoxin formation on sixteen soybean varieties. *J. Assoc, Off. Anal. Chem.,* 61, 574, 1978.
54. **Steele, J. L., Saul, R. A., and Hukill, W. V.,** Deterioration of shelled corn as measured by carbon dioxide production, *Trans. ASAE,* 12, 685, 1969.
55. **Stroshine, R. L., Kirleis, A. W., Tuite, J., Bauman, L. F., and Emam, A.,** Differences in grain quality among selected corn hybrids, *Cereal Foods World,* 31, 311, 1986.
56. **Stroshine, R., Tuite, J., Foster, G. H., and Baker, K.,** Gram drying and storage self-study guide, International Programs in Agriculture, Purdue University, Lafayette, Ind., 1985.
57. **Suryanarayana-Rao, K. and Tulpule, P. G.,** Varietal differences of groundnut in the production of aflatoxin, *Nature,* 214, 738, 1967.
58. **Tsubouchi, H., Yamamoto, K., Hisada, K., Sakabe, Y., and Tsuchihira, K.,** Inhibitory effects of non-aflatoxigenic fungi on aflatoxin production in rice cultivars by *Aspergillus flavus, Trans. Mycol. Soc. Japan,* 22, 103, 1981.
59. **Tsuruta, O., Gohara, S., and Saito, M.,** Scanning electron microscope observations of a fungal invasion of corn kernels, *Trans. Mycol. Soc. Japan,* 22, 121, 1981.
60. **Tuite, J. and Foster, G. H.,** Control of storage diseases of grain, *Annu. Rev. Phytopathol.,* 17, 343, 1979.
61. **Tuite, J., Koh-Knox, C., Stroshine, R., Cantone, R. A., and Bauman, L. F.,** Effect of physical damage to corn kernels on the development of *Penicillium* species and *Aspergillus glaucus* in storage, *Phytopathology,* 75, 1137, 1985.
62. **Wallin, J. R.,** Production of aflatoxin in wounded and whole maize kernels by *Aspergillus flavus, Plant Dis.,* 70, 429, 1986.
63. **Wicklow, D. T., Hesseltine, C. W., Shotwell, O. L., and Adams, G. L.,** Interference competition and aflatoxin levels in corn, *Phytopathology,* 70, 761, 1980.
64. **Widstrom, N. W., McMillian, W. W., Wilson, D. M., Garwood, D. L., and Glover, D. V.,** Growth characteristics of *Aspergillus flavus* on agar infused with maize kernel homogenates and aflatoxin contamination of whole kernel samples, *Phytopathology,* 74, 887, 1984.
65. **Widstrom, N. W., Wilson, D. M., and McMillian, W. W.,** Ear resistance of maize inbreds to field aflatoxin contamination, *Crop Sci.,* 24, 1155, 1984.
66. **Wilson, D. M., Mixon, A. C., and Troeger, J. M.,** Aflatoxin contamination of peanuts resistant to seed invasion by *Aspergillus flavus, Phytopathology,* 67, 922, 1977.
67. **Yao, B.,** personal communication, 1986.
68. **Yasue, T. and Kinomura, N.,** Studies on the mechanism of seed coat cracking and its prevention in soybeans. I. Regional and varietal difference in cracking of seed coat in soybeans, *Japan. J. Crop, Sci.,* 53, 93, 1984.

INDEX

A